Accounting, Finance, Sustainability, Governance & Fraud: Theory and Application

Series Editor

Kiymet Tunca Çalıyurt
Business Administration Department & Accounting and Finance Department
Trakya University
Edirne, Türkiye

This Scopus indexed series acts as a forum for book publications on current research arising from debates about key topics that have emerged from global economic crises during the past several years. The importance of governance and the will to deal with corruption, fraud, and bad practice, are themes featured in volumes published in the series. These topics are not only of concern to businesses and their investors, but also to governments and supranational organizations, such as the United Nations and the European Union. Accounting, Finance, Sustainability, Governance & Fraud: Theory and Application takes on a distinctive perspective to explore crucial issues that currently have little or no coverage. Thus the series integrates both theoretical developments and practical experiences to feature themes that are topical, or are deemed to become topical within a short time. The series welcomes interdisciplinary research covering the topics of accounting, auditing, governance, and fraud.

Ashish Virk • Aman Amrit Cheema
Pooja Sikka
Editors

Law, Economy, and Society: Insights into Legal and Economic Issues in India

Springer

Editors
Ashish Virk
University Institute of Laws
Panjab University Regional Centre
Ludhiana, Punjab, India

Aman Amrit Cheema
University Institute of Laws
Panjab University Regional Centre
Ludhiana, Punjab, India

Pooja Sikka
University Institute of Laws
Panjab University Regional Centre
Ludhiana, Punjab, India

ISSN 2509-7873 ISSN 2509-7881 (electronic)
Accounting, Finance, Sustainability, Governance & Fraud: Theory and Application
ISBN 978-981-95-3360-2 ISBN 978-981-95-3361-9 (eBook)
https://doi.org/10.1007/978-981-95-3361-9

This Springer imprint is published by the registered company Springer Nature Singapore Pte Ltd.
The registered company address is: 152 Beach Road, #21-01/04 Gateway East, Singapore 189721, Singapore

Contents

About the Editors

Ashish Virk is a distinguished academician and legal scholar renowned for her expertise in jurisprudence, human rights, and constitutional law. With over 20 years of teaching experience, she has contributed significantly to legal education at various institutions, including Panjab University Regional Centre, Ludhiana. As a prolific researcher, Prof. Virk has authored two books, contributed chapters to course work, and published 45 papers in international and national journals, six chapters in books, and five online publications and law magazine bulletins. She has presented 55 papers at conferences, including session chairing and paper presenting at Cambridge University, United Kingdom. She has also delivered 25 invited lectures, chaired and judged nine sessions and events, and participated in numerous radio talks on various legal topics. Her administrative roles include serving as a visiting faculty member at the Indian Institute of Management-Rohtak, as principal at Panjab University Constituent College, and as a lifetime member of the Indian Law Institute (Deemed University), New Delhi. Additionally, she has supervised Ph.D. theses and guided students on legal aid projects, demonstrating her commitment to academia and societal welfare. Prof. Virk's extensive academic contributions, administrative leadership, and commitment to student mentorship underscore her invaluable role in the legal fraternity.

Aman Amrit Cheema, a distinguished law luminary with over 22 years of experience as an educator and scholar. Recognized for her specialization in Narcotic Laws and the Juvenile Justice System, Dr. Cheema's academic journey is marked by significant milestones. With a Ph.D. from the University of Jammu, her research on the Narcotic Drugs and Psychotropic Substances Act has garnered acclaim for its socio-legal insights. She demonstrates exemplary leadership as Director and Professor of Laws at Panjab University Regional Centre, Ludhiana. Prof. Cheema's contributions transcend academia, including her honorary professorship at the International Institute of Crime and Security Sciences, Bengaluru. She has to her credit two books and more than 44 research publications. She has actively participated in 44 international and national conferences, including a paper presented and Session Chaired atthe International Conference on Law, Economics and Politics, Oxford University,

Oxford, United Kingdom (UK). She is a lifetime member of the Indian Law Institute (Deemed University), New Delhi. She has delivered radio talks on varied law topics on Prasar-Bharati, AIR FM Ludhiana, All India Radio-Jalandhar, and AIR FM Gold Ludhiana. She has been invited for 31 lectures, panel discussions, and keynote addresses and as a resource person at various eminent institutions. Her extensive administrative roles, editorial contributions, and membership in prestigious committees reflect her dedication to legal education and societal welfare. A prolific scholar and esteemed academician, Prof. Cheema continues to shape legal discourse through her expertise and commitment.

Pooja Sikka is an Assistant Professor in Economics at the University Institute of Laws, Panjab University Regional Centre, Ludhiana. She earned her Ph.D. in Economics from Punjabi University, Patiala, with a fellowship from the University Grants Commission. With 18 years of teaching experience, her research interests include Economic Growth, Development, Poverty, Income Inequality, and Public Policy. She has authored two books, including one by Routledge, and has 24 publications in esteemed journals, book chapters, and conference proceedings. Dr. Sikka actively participates in academic discourse, presenting papers at national and international conferences, including IGIDR Mumbai and CESS Hyderabad. She has demonstrated her expertise by delivering 10 engaging invited lectures, adeptly chairing sessions broadcasted on FM radio, and providing valuable insights as a resource person. She has effectively coordinated several national/ international academic conferences, seminars, extension lectures, faculty development programs, sensitivity programs, special drives and camps, and awareness campaigns in collaboration with prestigious institutions.

Part I
Introduction

Chapter 1
Legal and Economic Synergy for an Equitable Society

Ashish Virk, Aman Amrit Cheema, and Pooja Sikka

Abstract Integrating legal and economic frameworks is pivotal for fostering equity, addressing social disparities, and promoting justice in dynamic and interconnected societies. This chapter examines the critical interplay between these frameworks, emphasizing the role of legal structures in ensuring market efficiency, safeguarding property rights, and facilitating economic development. Drawing on theoretical perspectives, empirical evidence, and case studies, it explores how legal reforms such as land redistribution and progressive taxation address inequities and enhance social welfare. The analysis highlights the impact of regulatory mechanisms, including antitrust laws, in correcting market failures and fostering competition while acknowledging challenges like regulatory capture and the rapid evolution of digital economies. Social justice remains central to economic policy, with progressive taxation, social safety nets, and equitable education policies identified as key instruments for reducing inequalities and fostering social mobility. The study underscores the importance of adapting legal and economic frameworks to global integration and technological disruptions, emphasizing the need for innovation, international regulatory cooperation, and continuous policy refinement. By synthesizing diverse insights, this chapter comprehensively explains how a dynamic legal-economic synergy can promote sustainable development and inclusive growth in an increasingly complex world.

Keywords Legal-economic synergy · Social justice · Regulatory frameworks · Progressive taxation · Market efficiency

---

The chapters in this book are presented in the "International Conference on Legal and Economic Synergy for an Equitable Society, organised by the University Institute of Laws at Panjab University Regional Centre, Ludhiana, India on February 1st, 2024".

A. Virk (✉) · A. A. Cheema · P. Sikka
University Institute of Laws, Panjab University Regional Centre, Ludhiana, Punjab, India
e-mail: ashishvirk77@gmail.com; aamanrit@pu.ac.in; pooja24@pu.ac.in

A. Virk et al. (eds.), *Law, Economy, and Society: Insights into Legal and Economic Issues in India*, Accounting, Finance, Sustainability, Governance & Fraud: Theory and Application, https://doi.org/10.1007/978-981-95-3361-9_1

1.1 Introduction

Pursuing an equitable society necessitates a harmonious blend of legal and economic frameworks. This synergy is fundamental in addressing social disparities, promoting justice, and ensuring sustainable development. Law is the backbone of societal organization, providing the necessary structure for economic activities. Effective legal frameworks ensure property rights, enforce contracts, and regulate market activities, thereby creating a conducive environment for economic development. According to Acemoglu and Robinson (2012), countries with strong legal institutions tend to have higher levels of economic prosperity compared to those with weaker legal systems. This foundation allows for the exploration of how legal frameworks contribute to economic equity.

Building on the idea that law underpins economic activities, legal reforms play a pivotal role in addressing economic inequities. Land reforms, for instance, have been instrumental in redistributing wealth and reducing poverty in several developing countries. The Indian experience with land reforms, as discussed by Besley and Burgess (2000), highlights the positive impact of legal interventions on economic equity. These reforms demonstrate how targeted legal changes can promote fairness and reduce disparities in wealth distribution.

Economic theories provide a critical lens through which legal frameworks can be evaluated and designed. The concept of Pareto efficiency, for example, emphasizes the importance of resource allocation where no individual can be made better off without making someone else worse off. This principle is crucial in legal decision-making, particularly in areas such as taxation and welfare distribution (Arrow 1963). By applying economic theories, legal structures can be designed to achieve optimal resource distribution and enhance societal welfare.

Regulatory mechanisms are essential in correcting market failures and ensuring fair competition. Antitrust laws, for instance, prevent monopolies and promote market efficiency, thereby protecting consumer interests. Stigler (1971) argues that regulatory policies must be designed to minimize regulatory capture and ensure that they serve the public interest rather than special interest groups. This underscores the need for effective legal frameworks that align with economic principles to maintain competitive and fair markets.

The concept of social justice is deeply intertwined with economic policies. Redistribution of wealth through progressive taxation and social welfare programs is a legal mechanism aimed at reducing income inequality. Piketty (2014) emphasizes the need for robust legal frameworks to address the increasing concentration of wealth and ensure a more equitable distribution of resources. These legal mechanisms are crucial in creating a fair society where economic benefits are shared more evenly across the population.

The advent of digital technologies has transformed economic landscapes, necessitating new legal frameworks to address emerging challenges. The regulation of digital markets, data protection, and cybersecurity are critical areas where law and economics intersect. Brynjolfsson and McAfee (2014) highlight the need for adaptive legal structures to harness the benefits of digital advancements while mitigating associated risks. This adaptation is necessary to ensure that the legal system keeps pace with technological changes and continues to promote equity.

Different countries have adopted various approaches to integrate legal and economic policies to promote equity. The Scandinavian model, characterized by strong legal protections and comprehensive welfare systems, exemplifies how legal and economic synergies can create more equitable societies. Andersen et al. (2007) discuss how the integration of social policies with economic strategies in Scandinavian countries has led to high levels of social equity and economic efficiency. This model illustrates the potential of coordinated legal and economic policies in achieving societal goals.

Despite the clear benefits of legal and economic synergy, several challenges persist. Legal pluralism, where multiple legal systems coexist, can create inconsistencies and hinder economic development. Additionally, global economic integration poses challenges to national legal systems, requiring harmonization of laws across borders (Rodrik 2011). These challenges highlight the need for continuous refinement of legal frameworks to respond to evolving economic and social environments.

The synergy between legal and economic frameworks is crucial for fostering an equitable society. By ensuring fair resource distribution, promoting justice, and facilitating sustainable development, this synergy addresses the root causes of social disparities. The integration of legal and economic policies must be continuously refined to adapt to changing societal needs and global economic trends. Drawing from a rich body of scholarly work, this interdisciplinary approach is essential in achieving social equity.

The objective of this chapter is to analyze the interplay between legal structures and economic mechanisms in promoting equity. By examining various theoretical perspectives, legal reforms, and regulatory practices, the chapter seeks to highlight the importance of an integrated approach to law and economics. The analysis will also consider the impact of digital technologies on these frameworks and propose adaptive strategies for future policy development. The methodology involves a thorough literature review of scholarly works, focusing on studies that explore the relationship between law and economics. This includes analyzing empirical research, theoretical models, and case studies from different countries. The chapter synthesizes these insights to provide a holistic understanding of the legal and economic synergy necessary for achieving social equity. In this chapter, three key issues will be discussed in the subsequent sections: the role of legal frameworks in economic development, the impact of regulatory mechanisms on market efficiency, and the importance of social justice in economic policies. These discussions aim to provide a comprehensive understanding of how legal and economic synergies can lead to a more equitable society.

1.2 Role of Legal Frameworks in Economic Development

Legal frameworks are fundamental to economic development, providing the rules and institutions that facilitate market transactions, protect property rights, enforce contracts, and ensure fair and efficient economic functioning.

This section explores the multifaceted impact of legal frameworks on economic development, drawing on theoretical perspectives, empirical evidence, and case studies from various countries.

1.2.1 Theoretical Perspectives on Legal Frameworks and Economic Development

The New Institutional Economics (NIE) theory, advanced by Douglass North, posits that institutions, including legal frameworks, are fundamental determinants of economic performance (North 1990). According to NIE, institutions reduce transaction costs, mitigate risks, and provide the stability necessary for economic actors to engage in productive activities. Property rights are central to this theoretical framework. Secure property rights incentivize individuals and businesses to invest in assets, leading to capital accumulation and economic growth. The assurance that one's property will not be expropriated or encroached upon without due process encourages investment in land, buildings, machinery, and intellectual property. This, in turn, fosters innovation, entrepreneurship, and economic expansion.

Contract enforcement is another critical aspect of legal frameworks. Efficient contract enforcement mechanisms ensure that agreements between parties are honored, reducing the uncertainty and risks associated with economic transactions. This predictability encourages businesses to engage in long-term planning and investments, knowing that their contractual rights will be protected by the legal system. Regulatory policies that promote fair competition and protect consumer rights are essential for economic development. Effective regulations prevent market abuses, monopolistic practices, and exploitation, fostering a competitive market environment that drives efficiency and innovation. Conversely, weak or inconsistent regulatory frameworks can lead to market failures, corruption, and economic inefficiencies.

1.2.2 Empirical Evidence on Legal Frameworks and Economic Development

Empirical studies provide robust evidence supporting the theoretical link between legal frameworks and economic development. La Porta et al. (1998) examined the legal systems of 49 countries and their impact on financial development. The authors found that countries with stronger legal protections for investors, as measured by the quality of their legal frameworks, had more developed financial markets. This finding underscores the importance of legal institutions in facilitating access to finance, which is crucial for economic growth. Acemoglu and Johnson (2005) explored the role of property rights and contract enforcement in economic development. Using historical data and cross-country comparisons, the authors concluded that property rights institutions have a significant impact on economic performance. Countries with secure property rights experienced higher levels of investment and economic growth, while those with weak property rights faced persistent economic challenges.

The World Bank's Doing Business reports also provide valuable insights into the relationship between legal frameworks and economic development. These reports assess the ease of doing business in various countries by examining factors such as property registration, contract enforcement, and regulatory quality. Countries that rank higher in these indicators tend to have more dynamic and prosperous economies. For instance, Singapore, which consistently ranks among the top in the Doing Business reports, has a well-developed legal system that supports business activities and economic growth (World Bank 2020).

Several case studies illustrate how legal reforms have contributed to economic development in different contexts. One notable example is the legal reforms in China since the late 1970s. The introduction of property rights and the establishment of a market-oriented legal framework were crucial components of China's economic transformation. These reforms provided the legal certainty needed for domestic and foreign investors to engage in economic activities, leading to rapid industrialization and growth (Zhang 2019).

Another compelling case is the legal reforms in Eastern Europe following the collapse of the Soviet Union. Countries like Poland, Hungary, and the Czech Republic undertook comprehensive legal reforms to transition from centrally planned economies to market economies. These reforms included the establishment of private property rights, the creation of commercial laws, and the development of regulatory frameworks to support market activities. As a result, these countries experienced significant economic growth and integration into the global economy (EBRD 2004).

In Sub-Saharan Africa, Rwanda's legal reforms provide a noteworthy example of how improving the legal framework can spur economic development. Over the past two decades, Rwanda has implemented a series of legal and institutional reforms aimed at enhancing the business environment. These reforms include simplifying property registration, improving contract enforcement, and reducing regulatory burdens. Consequently, Rwanda has seen substantial improvements in its Doing Business rankings and has become one of the fastest-growing economies in the region (World Bank 2020).

Therefore, legal frameworks are foundational to economic development, providing the rules and institutions that facilitate market transactions, protect property rights, and ensure the efficient functioning of the economy.

1.3 The Impact of Regulatory Mechanisms on Market Efficiency

Regulatory mechanisms are crucial in shaping market efficiency by setting and enforcing rules that address market failures, foster competition, and protect consumer interests. These mechanisms help maintain a balanced marketplace where businesses can compete fairly, and consumers benefit from improved services and lower prices.

In the United States, antitrust laws play a central role in regulating market efficiency. The Sherman Act of 1890 was among the earliest legislation aimed at combating monopolistic practices and promoting competition by prohibiting restraints on trade and commerce. This Act was expanded with the Clayton Act and the Federal Trade Commission Act, which targeted issues such as price discrimination and unfair competition (Hovenkamp 2004). A notable example of the impact of these antitrust regulations is the breakup of AT&T in 1982. This action dismantled a monopoly and spurred competition in the telecommunications sector, resulting in lower prices and increased innovation. This case highlights the effectiveness of antitrust laws in improving market efficiency by addressing and correcting monopolistic structures (Hovenkamp 2004).

Similarly, the European Union has implemented a robust competition policy framework that significantly enhances market dynamics. The EU's competition policy includes rigorous enforcement of rules on mergers and acquisitions and prevention of anti-competitive practices (Whish and Bailey 2018). For example, the EU's decision to block the Siemens-Alstom merger in 2019 was driven by concerns that the merger would reduce competition and harm consumer interests. This decision underscores the EU's commitment to maintaining a competitive market environment and illustrates the effectiveness of its competition policies (Crisp 2019).

However, regulatory mechanisms face several challenges that can impact their effectiveness. One significant issue is regulatory capture, where industries may exert undue influence on regulators to shape policies in their favor. This situation can undermine the effectiveness of regulations and lead to market distortions that negatively impact consumers and stifle competition (Stigler 1971). Addressing regulatory capture requires enhancing transparency and accountability in the regulatory process to ensure that regulations serve the public interest rather than specific industry interests.

Another challenge is the need for regulatory frameworks to adapt to rapidly changing market conditions, especially in the digital economy. The rise of digital platforms and e-commerce has introduced new regulatory challenges that traditional frameworks may not adequately address. For instance, the growth of digital platforms has raised concerns about market dominance, data privacy, and consumer protection. Regulators must develop adaptive strategies to manage these new challenges and ensure that regulations remain effective in promoting market efficiency.

The increasing complexity of global markets also poses a challenge for regulatory frameworks. As businesses operate across borders, there is a growing need for international cooperation and harmonization of regulations to address issues that transcend national boundaries. This includes aligning standards for digital privacy, intellectual property, and competition policies to prevent regulatory arbitrage and ensure a level playing field globally (Gans and Stern 2021). Collaborative efforts between international regulatory bodies can help create cohesive frameworks that facilitate global trade while protecting consumer interests and promoting fair competition.

Finally, the role of technology in regulatory innovation cannot be overlooked. Emerging technologies such as artificial intelligence and blockchain have the potential to transform regulatory practices by enhancing data collection, enforcement capabilities, and transparency. For example, blockchain technology can improve the traceability and security of transactions, reducing fraud and increasing trust in regulatory processes (Catalini and Gans 2016). Embracing technological advancements and integrating them into regulatory frameworks can lead to more efficient and effective governance, ultimately fostering a more equitable and competitive marketplace.

Thus, regulatory mechanisms play a crucial role in enhancing market efficiency by promoting fair competition, protecting consumers, and addressing market failures. While there have been notable successes in the United States and the European Union, challenges such as regulatory capture, the need for adaptive strategies in the digital economy, global market complexity, and the integration of new technologies persist.

1.4 Role of Social Justice in Economic Policy Frameworks

Social justice is a vital principle that shapes how economic policies are formulated and implemented, ensuring that fairness and equity guide the distribution of resources and opportunities within society. This principle is crucial for addressing systemic inequalities that can arise from various socio-economic factors such as income disparity, educational access, and healthcare availability.

A foundational element of promoting social justice through economic policies is progressive taxation. Progressive tax systems are designed so that individuals with higher incomes contribute a larger percentage of their earnings in taxes. This approach aims to reduce income inequality by redistributing wealth from the more affluent to those with fewer resources. Progressive taxation not only helps to level the economic playing field but also funds essential public services, including education, healthcare, and social welfare programs. By doing so, it supports economic opportunities for all individuals, particularly those who are disadvantaged (Piketty 2014).

In addition to progressive taxation, social safety nets play a crucial role in promoting social justice. Programs such as unemployment benefits, social security, and universal healthcare are designed to provide support to those in need and mitigate the impacts of economic downturns. These safety nets help reduce poverty and ensure that everyone has access to basic necessities, contributing to a more equitable society where individuals have the opportunity to thrive (Esping-Andersen 1990).

Education policies also significantly impact social justice. Access to quality education is essential for reducing socio-economic disparities and equipping individuals with the skills necessary for full economic participation. Policies that focus on improving educational outcomes for disadvantaged communities—such as increasing funding for public schools in low-income areas and providing scholarships for higher education—are crucial for creating equal opportunities and fostering social mobility (Checchi 2006). By investing in education, societies can help break the cycle of poverty and promote long-term economic equity.

Labor market policies further contribute to social justice by ensuring fair wages, job security, and safe working conditions. Minimum wage laws and labor rights protections are designed to prevent exploitation and ensure that workers are compensated fairly for their labor. These policies help create a more just economic system by supporting fair labor practices and improving working conditions (Katz and Krueger 2016). Such measures are vital for ensuring that economic growth translates into tangible benefits for workers and their families.

Moreover, addressing systemic discrimination is a crucial aspect of promoting social justice. Economic policies should ensure equal opportunities regardless of race, gender, or other factors. Affirmative action programs, anti-discrimination laws, and efforts to close the gender pay gap are examples of how policies can work to eliminate barriers and promote equity in the workplace and beyond (Blau and Kahn 2017). By tackling these issues, societies can work towards overcoming historical injustices and creating a more inclusive economy where everyone has the opportunity to succeed.

Despite these efforts, challenges remain in achieving social justice through economic policies. Structural inequalities and entrenched power dynamics can hinder progress towards a more equitable society. For example, economic policies may sometimes disproportionately benefit certain groups while neglecting others, particularly when the interests of powerful stakeholders overshadow the needs of marginalized populations. Addressing these challenges requires ongoing monitoring and evaluation of economic policies to ensure they are inclusive and responsive to the needs of all individuals (Sen 1999).

International organizations also play a critical role in advancing social justice on a global scale. The United Nations Development Programme (UNDP) and other international bodies work towards promoting equitable economic development and reducing global poverty. Their initiatives often focus on supporting developing countries, addressing global inequalities, and promoting fair trade practices. Through international cooperation and support, these organizations help foster social justice globally, ensuring that economic benefits are shared more equitably across nations (UNDP 2020).

In the era of rapid technological advancement, ensuring that economic policies incorporate social justice remains essential. Digital technologies have the potential to both create opportunities and exacerbate inequalities. To mitigate these disparities, policies must ensure equitable access to technology and digital literacy programs, helping to close the digital divide and promote inclusive economic development (OECD 2021).

Therefore, social justice is integral to economic policies aimed at creating a fair and equitable society. By implementing progressive taxation, providing robust social safety nets, investing in education, ensuring fair labor practices, and addressing systemic discrimination, societies can work towards greater economic equity. Despite the challenges, ongoing commitment to monitoring and adapting policies, as well as international cooperation, is crucial for advancing social justice and ensuring that economic growth benefits everyone.

1.5 Concluding Remarks

The interplay between legal and economic frameworks is essential for an equitable society. Legal structures underpin economic activities by ensuring efficient markets, contract enforcement, and property rights protection. Historical and contemporary reforms, such as land redistribution and progressive taxation, have shown the potential to address disparities and promote fairness. Economic theories like Pareto efficiency, applied in legal design, optimize resource distribution and enhance societal welfare. Regulatory mechanisms correct market failures, promote fair competition, and safeguard consumer interests. The experiences of the United States and the European Union demonstrate the effectiveness of robust antitrust laws. However, regulatory capture and the rapid evolution of digital economies require adaptive strategies to remain effective.

Social justice must guide economic policies to ensure equitable resource and opportunity distribution. Progressive taxation, social safety nets, and education policies foster social mobility. Labor market policies and efforts to address systemic discrimination are also crucial. Despite progress, continuous refinement of legal and economic policies is necessary due to global integration and technological advancements. Strengthening property rights protections, enforcing antitrust laws, enhancing transparency to combat regulatory capture, and adapting regulations to digital challenges are all essential steps. Comprehensive social safety nets and investment in education for disadvantaged communities are vital for promoting social mobility. Further research should explore the impact of emerging technologies, international regulatory cooperation, and the long-term outcomes of legal reforms on economic equity. A dynamic and responsive legal-economic synergy is vital for addressing social disparities and ensuring economic growth benefits all members of society.

References

Acemoglu D, Johnson S (2005) Unbundling Institutions. J Polit Econ 113(5):949–995. https://doi.org/10.1086/432166

Acemoglu D, Robinson JA (2012) Why nations fail: the origins of power, prosperity, and poverty. Crown Publishers
Andersen TM, Holmström B, Honkapohja S, Korkman S, Söderström HT, Vartiainen J (2007) The Nordic model: embracing globalization and sharing risks. The Research Institute of the Finnish Economy (ETLA)
Arrow KJ (1963) Social choice and individual values. Yale University Press
Besley T, Burgess R (2000) Land reform, poverty reduction, and growth: evidence from India. Q J Econ 115(2):389–430. https://doi.org/10.1162/003355300554809
Blau FD, Kahn LM (2017) The gender pay gap: a review of the evidence and policy options. ILR Rev 70(3):727–757. https://doi.org/10.2202/1553-3832.1190
Brynjolfsson E, McAfee A (2014) The second machine age: work, progress, and prosperity in a time of brilliant technologies. W.W. Norton & Company
Catalini C, Gans JS (2016) Some simple economics of the Blockchain. Commun ACM 59(11):32–34. https://doi.org/10.2139/ssrn.2874598
Checchi D (2006) The economics of education: human capital, family background and inequality. Cambridge University Press
Crisp J (2019) EU blocks Siemens-Alstom merger over competition concerns. Financial Times. Retrieved from https://www.ft.com/content/6e344f6a-29fd-11e9-88a4-c32129756dd8
EBRD (2004) Transition report 2004: infrastructure. European Bank for Reconstruction and Development
Esping-Andersen G (1990) The three worlds of welfare capitalism. Princeton University Press
Gans JS, Stern S (2021) The economics of innovation: an introduction. J Econ Perspect 35(3):3–25. https://doi.org/10.1257/jep.35.3.3
Hovenkamp H (2004) Federal antitrust policy: the law of competition and its practice. West Academic Publishing
Katz LF, Krueger AB (2016) The rise and nature of alternative work arrangements in the United States, 1995-2015. Brook Pap Econ Act 2016(2):289–337. https://doi.org/10.1177/0019793918820008
La Porta R, Lopez-de-Silanes F, Shleifer A, Vishny R (1998) Law and finance. J Polit Econ 106(6):1113–1155. https://doi.org/10.1086/250042
North DC (1990) Institutions, institutional change and economic performance. Cambridge University Press
OECD (2021) Digital economy outlook 2021. Organisation for Economic Co-operation and Development. Retrieved from https://www.oecd.org/digital/digital-economy-outlook. Accessed 15 Mar 2024
Piketty T (2014) Capital in the twenty-first century. Harvard University Press
Rodrik D (2011) The globalization paradox: democracy and the future of the world economy. W.W. Norton & Company
Sen A (1999) Development as freedom. Oxford University Press
Stigler GJ (1971) The theory of economic regulation. Bell J Econ Manag Sci 2(1):3–21. https://doi.org/10.2307/3003160
UNDP (2020) Human development report 2020: the next frontier—human development and the Anthropocene. United Nations Development Programme. http://hdr.undp.org/en/2020-report. Accessed 15 Mar 2024
Whish R, Bailey D (2018) Competition law. Oxford University Press
World Bank (2020) Doing business 2020: comparing business regulation in 190 economies. The World Bank
Zhang W (2019) Legal reforms and economic development in China. Asian Econ Policy Review 14(1):50–68. https://doi.org/10.1111/aepr.12239

Prof. (Dr.) Ashish Virk is a distinguished academic and legal scholar renowned for her expertise in jurisprudence, human rights, and constitutional law. With over 20 years of teaching experience, she has made significant contributions to legal education at various institutions, including the Panjab University Regional Centre in Ludhiana. As a prolific researcher, Prof. Virk has authored two books, contributed chapters to coursework, and published 45 papers in international and national journals, six chapters in books, and five online publications, as well as articles in law magazine bulletins. She has presented 55 papers at conferences, including session chairing and paper presenting at Cambridge University, UK. She has also delivered 25 invited lectures, chaired and judged nine sessions and events, and participated in numerous radio talks on various legal topics. Her administrative roles include serving as a visiting faculty member at the Indian Institute of Management—Rohtak, as principal at Panjab University Constituent College, and as a lifetime member of the Indian Law Institute (Deemed University), New Delhi. Additionally, she has supervised Ph.D. theses and guided students on legal aid projects, demonstrating her commitment to academia and societal welfare. Prof. Virk's extensive academic contributions, administrative leadership, and commitment to student mentorship underscore her invaluable role in the legal fraternity.

Prof. (Dr.) Aman Amrit Cheema is a distinguished law luminary with over 22 years of experience as an educator and scholar. Recognized for her specialization in Narcotic Laws and the Juvenile Justice System, Dr. Cheema's academic journey is marked by significant milestones. With a Ph.D. from the University of Jammu, her research on the Narcotic Drugs and Psychotropic Substances Act has garnered acclaim for its socio-legal insights. She demonstrates exemplary leadership as the Director and Professor of Law at the Panjab University Regional Centre, Ludhiana. Prof. Cheema's contributions extend beyond academia, including her honorary professorship at the International Institute of Crime and Security Sciences in Bengaluru. She has to her credit two books and more than 44 research publications. She has actively participated in 44 international and national conferences, including presenting a paper and chairing a session at the International Conference on Law, Economics, and Politics at Oxford University, Oxford, UK. She is a lifetime member of the Indian Law Institute (Deemed University), New Delhi. She has delivered radio talks on various law topics on Prasar Bharati, AIR FM Ludhiana, All India Radio Jalandhar, and AIR FM Gold Ludhiana. She has been invited to deliver 31 lectures, panel discussions, and keynote addresses and has also served as a resource person at various esteemed institutions. Her extensive administrative roles, editorial contributions, and membership in prestigious committees reflect her dedication to legal education and societal welfare. A prolific scholar and esteemed academician, Prof. Cheema continues to shape legal discourse through her expertise and commitment.

Dr. Pooja Sikka is Assistant Professor of Economics at the University Institute of Laws, Panjab University Regional Centre, Ludhiana. She earned her Ph.D. in Economics from Punjabi University, Patiala, with a fellowship from the University Grants Commission of India. With 18 years of teaching experience, her research interests include economic growth, development, poverty, income inequality, and public policy. She has authored two books, including one by Routledge, and has 24 publications in esteemed journals, book chapters, and conference proceedings. Dr. Sikka actively participates in academic discourse, presenting papers at national and international conferences, including the IGIDR in Mumbai and CESS in Hyderabad. She has demonstrated her expertise by delivering 10 engaging invited lectures, adeptly chairing sessions broadcast on FM radio, and providing valuable insights as a resource person. She has effectively coordinated several national/international academic conferences, seminars, extension lectures, faculty development programmes, sensitivity programmes, special drives and camps, and awareness campaigns in collaboration with prestigious institutions.

Prof. (Dr.) [illegible] [illegible] [illegible] [illegible] [illegible] [illegible] [illegible] She demonstrates exemplary [illegible] as the Director and Professor of Law at the Panjab University Regional Centre, Ludhiana. [illegible] [illegible] [illegible] [illegible] [illegible] [illegible] [illegible] [illegible] [illegible] [illegible] [illegible] [illegible] [illegible] [illegible]

Dr. Pooja Sikka [illegible] [illegible] [illegible] [illegible] [illegible] [illegible] [illegible] [illegible] [illegible]

Part II
Legal Issues and Economic Perspective

Chapter 2
The Digital Dilemma: Regulating Social Media in India

Homa Bansal and Jesus Goyal

Abstract The rise of social media has transformed the way people communicate, creating an active and interactive platform where individuals can share their views, exchange ideas, and engage in discussions worldwide. As an essential part of digital life, social media helps in democratic participation, promotes transparency, and gives a voice to marginalized groups. In India, the right to freedom of speech and expression is guaranteed under Article 19(1)(a) of the Constitution, enabling individuals to express their opinions through various means, including online platforms. However, this right comes with reasonable restrictions under Article 19(2), which are meant to maintain public order, security, and morality. With over 600 million internet users, India's digital landscape has grown rapidly, making social media a powerful tool for public discourse. While it plays a key role in social change and political activism, there are serious concerns about misinformation, hate speech, cybercrimes, and government censorship. The legal and ethical issues related to online speech necessitate a balanced approach to regulation, ensuring that restrictions do not infringe upon fundamental rights. This study examines the evolving nature of free speech on social media in India, the challenges of content moderation, and the legal framework governing digital expression, while promoting responsible and democratic use of online platforms.

Keywords Government censorship · Social media · Freedom of speech · Transparency · Legal framework

The chapters in this books are presented in the "*International Conference on Legal and Economic Synergy for an Equitable Society, organised by the University Institute of Laws at Panjab University Regional Centre, Ludhiana, India on February 1st, 2024*".

H. Bansal (✉) · J. Goyal
University Institute of Laws, Panjab University Regional Centre, Ludhiana, Punjab, India
e-mail: homa.bansal57@gmail.com; jesusgoyal@gmail.com

A. Virk et al. (eds.), *Law, Economy, and Society: Insights into Legal and Economic Issues in India*, Accounting, Finance, Sustainability, Governance & Fraud: Theory and Application, https://doi.org/10.1007/978-981-95-3361-9_2

2.1 Introduction

The rise of the internet has transformed communication, with online interaction becoming a significant form of digital communication. Online entertainment, a dominant type of online activity, has become an integral part of people's lives, significantly influencing how they spend their time and shaping popular opinion. In India, where nearly half the population engages with online entertainment, its popularity has soared, especially during the COVID-19 pandemic. However, this widespread use comes with challenges. Online entertainment platforms are increasingly becoming channels for illegal activities. For instance, in November 2012, a young woman was arrested for a Facebook post expressing her views on the city's reaction to the death of Shiv Sena leader Bal Thackeray. The post questioned why the city shuts down after the death of a politician. Both she and a friend who "liked" the post were arrested, causing a public outcry. The court later sentenced them both (Parmar 2012).

2.2 Right to Speak on Social Media

Definition of Social Media
Social media is a computer-based system that facilitates the interchange of ideas, thoughts, and information through virtual collaboration. Andreas Kaplan and Michael Haenlein stated that it "consists of a set of Internet-based applications built on the ideas and technologies of Web 2.0 that enable the creation and sharing of user-generated content."

Usage Statistics in India
India's social media landscape is thriving, with citizens dedicating a significant portion of their day, around 2.25 h, to these platforms. This surge in internet usage is reflected in the country's rapidly growing online population, which now stands at an impressive six million users. This figure represents a substantial penetration rate, with nearly half (48.7%) of the total population actively engaging with the internet (Kemp 2023).

The Right to Speak Freely of Discourse and Articulation in India
In its role as a vote-based country, India maintains principal privileges, including the freedom to express one's feelings freely. Article 19(1)(a) of the Constitution of the Indian Republic ensures each resident the right to offer their viewpoints through any medium, including web-based opinion forums. While this opportunity is dependent upon lawful restrictions under Article 19(2), it remains a foundation of individual freedom and assumes a pivotal part in molding public talk. Occasions, for example, the Facebook post contention in November 2012 and the Baduria communal disruption in July 2017, underscore the challenges of balancing the use of this right enshrined under Article 19(1)(a) on social media (Das 2017). These

incidents highlight the imperative need for well-crafted laws to regulate social media, ensuring responsible use without compromising fundamental rights and fostering a democratic online environment (Raj 2008).

Navigating Internet Access, Freedom of Speech, and Social Media

In today's modern world, the internet has become a basic need, especially with the impact of the Coronavirus pandemic. Social networks played a crucial role in communication, supporting freedom of speech, and serving as a platform for global movements, fostering instant connections and solidarity.

The United Nations' Special Rapporteur highlights the significance of the cyber world in exercising freedom of speech, considering it a fundamental human right. Countries are strongly encouraged to ensure uninterrupted internet access even during times of political turmoil, emphasizing the positive impact of making it widely available, accessible, and affordable. This recognition is in line with international agreements like the Universal Declaration of Human Rights (UDHR) and the International Covenant on Civil and Political Rights (ICCPR), which affirm freedom of speech and its extension to the internet and social media (United Nations General Assembly 2011).

Recent legal decisions in India acknowledge the internet's essential role in daily life, considering it a key part of the framework for the rights enshrined under Article 19(1)(a). Nevertheless, freedom of speech isn't without its limits. Both international agreements and Indian laws acknowledge restrictions aimed at safeguarding the rights of individuals, as well as their reputation, public safety, order, and welfare.

In this evolving digital landscape, where information is king, governments worldwide grapple with the power of social media to spread information and expose wrongdoing. The internet's potential for good is matched by its potential for misuse, highlighting the need for sensible, unofficial rules. The challenge lies in regulating it without restricting basic rights, especiallyin India, which is classified as "Partly Free" regarding internet access, faces a surge in government requests to remove content. Changes to the Information Technology Act and the developing digital regulation scene reflect attempts to manage online content without compromising essential freedoms. Achieving a delicate balance requires guidelines that safeguard individual rights without stifling the diversity of opinions crucial for a healthy democracy.

Government efforts to require social media platforms to pre-approve content have sparked debates about free speech. The goal is to prevent harmful content, but there are concerns about potential censorship. Finding a balance requires rules that protect individual rights without stifling the variety of opinions necessary for a healthy democracy. The evolving landscape of internet access, free speech, and social media regulation requires a thoughtful and delicate balance. We need systems that safeguard individual rights while addressing the challenges of this networking era. While India doesn't have specific laws governing social media, existing digital regulations offer some tools to deal with online copyright infringement, including on the web and online networking platforms.

Today's world is defined by the omnipresence of data, which plays a vital role in our progress. The technological advancements that connect us in a fast-paced global network emphasize the value of data in driving change. Social media and citizen journalism rely on transparency and the sharing of information. Social media, as a powerful tool, has empowered everyday people to act as watchdogs, holding authorities accountable and exposing wrongdoing.

In the past, governments often tried to control the flow of information to the public. However, the rise of social media, with its powerful ability to spread information to everyone, is seen as a threat by governments that want to control it. The internet, as the foundation of modern development, plays a vital role in democracies worldwide, allowing citizens to participate beyond geographical limitations. However, governments, recognizing both the positive and negative potential of the internet, are inclined to regulate it. Despite its benefits, Cyberspace is vulnerable to misuse. This vulnerability justifies some government regulation of online content, driven by concerns for public safety. Social media platforms can become breeding grounds for various cybercrimes, including defamation, hacking, hate speech, stalking, and incitement to violence. Once harmful content is uploaded, its viral nature makes it difficult to contain. Therefore, some government regulation of social media seems necessary.

While government regulation is acceptable as long as it safeguards individual and collective interests, concerns arise when it transforms into censorship (Fisher 2012), encroaching upon civil rights such as freedom of speech and expression. Shields exist, yet states frequently outperform them to fluctuating degrees. China, strikingly, leads in web oversight with its unpredictable framework known as the "Incomparable Firewall of China" or "Brilliant Safeguard Task," regularly obstructing pages with frightful substance. In India, the situation with web opportunity is considered "Somewhat Free," as per Opportunity House's report. The public authority's rising command over the web places India under the class of "Nations under Reconnaissance." Google's straightforwardness report features a flood in Indian government demands for content expulsion, stressing the developing requirement for administrative measures.

The Information Technology Act, in force starting around 2000, got some forward movement following the 2008 Mumbai psychological militant assaults. Resulting changes extended the public authority's checking and blue-penciling abilities, putting India in the group of countries executing 'particular' web separating. In 2011, the public authority made a striking stride by encouraging web giants like Google, Facebook, and Microsoft to pre-screen information before web-based posting. The essential objective was to forestall hostile material as opposed to confine the right to speak freely of discourse. Be that as it may, this drive confronted reaction because of the difficulty of overseeing client produced content on such a tremendous scale.

In a landmark case, the High Court highlighted the essential job of a pluralistic and different media climate in maintaining the free discourse privileges of residents. The court emphasized that a successful democracy necessitates a variety of opinions and views on public issues, challenging any monopolistic control, whether by the state or any other entity. In conclusion, advocating for the regulation of social media rather than outright censorship appears to be a more judicious approach (Gupta 2012). This strategy aims to strike a delicate balance that protects users' rights while simultaneously safeguarding the interests of potential victims. This discussion naturally leads to an exploration of India's cyber laws, which indirectly target the regulation of social media.

2.3 Section 66A of the Information Technology Act, 2000

Among all the provisions provided in the Act of 2000, Section 66A has caused a lot of controversy in recent years. The Information Technology (Amendment) Act, 2008 inserted Section 66A into the Information Technology Act, which provides penalties for sending objectionable information through communication and provides:

Any person, through a computer or communication device, sends:—any material that is offensive or threatening; The use of any computer information or communications device, any attempt to cause distress or inconvenience to, or deceive or defraud, any resident or recipient shall constitute the basis hereof. The purpose of e-mail or e-mail messages is punishable by imprisonment for up to three years and a fine.

2.4 Explanation

For this section, the terms "e-mail" and "e-mail message" mean a message created, sent, or received on a computer or communications equipment (text, image, voice, video) or refers to messages. video and other electronic data that may be transmitted in messages. Looking at this provision, it is clear that there is a contradiction between the words of Section 66A of the IT Act and Article 19(1)(a), which guarantees the right of all Indians to freedom of speech and expression (Anuradha Basin v. Union of India 2020). Article 19(2) distinctly outlines the grounds for reasonable limitations on freedom of speech and expression. However, under section 66A, restrictions on freedom of speech and expression are limited to many things other than those specified in the constitution. There are some flaws in this provision that interfere with the right to freedom of expression. Words like "violent," "threatening," "obstructive," "dangerous," "obstructive," "derogatory," and "hurtful" have no literal meaning.

Many cases of misuse of Section 66A have been reported across the country, adding to the controversy. In November 2012, A businessman from Puducherry, Mr. Ravi Srinivasan, was arrested under Section 66A for accusing the then Finance Minister, Mr. P. Chidambaram's son, Karti Chidambaram of corruption on Twitter. Similarly, in September 2012, independent actor Aseem Trivadi was also arrested under Section 66A of the IT Act and other sections of the IPC and the Prevention of Crimes Against National Honor Act, 1971. He was arrested because his paintings depicting widespread corruption in India were considered terrorism. Additionally, two Air India employees were arrested by Mumbai police the same year for sharing content on Facebook criticizing union leaders and some politicians; They were detained for 12 days. These events attracted great public attention.

Following this, numerous petitions inundated various courts in India, contesting the constitutionality of Section 66A. In 2012, Shreya Singhal, a law student from Delhi, filed a PIL in the Supreme Court of India (Shreya Singhal v. Union of India 2015). Her petition challenged the constitutionality of the law, arguing that its vagueness violated Articles 14, 19(1)(a) and 21 of the Constitution. Similar petitions were filed by the founders of MouthShut.com and the NGO Common Cause. It was accepted on 29-11-2012, and on 24-03-2015, the Apex Court ruled Section 66A of the Information Technology Act, 2000 as completely invalid. The court deemed it "illegal, excessive, and incompatible with the right to freedom of expression under Article 19(1) of the Constitution of India".

2.5 Current Scenario

With the popularity of the internet, the number of social media users in India reached 440 million in January 2021, and the user base is expected to reach around 1 billion or more by 2040. This increase underscores the urgent need for social governance. Social media space. The media's unique role in the democratic world is, of course, to bring citizens together across borders and voice their concerns. Even if the body cannot do this, the resistance power is still weak (Kamath 2020).

A recent incident in Kerala exemplifies the positive changes brought about by social media. One and a half-year-old Mohammad suffers from Spinal Muscular Atrophy (SMA) and needs the expensive medicine Zolgensma, which costs Rs 18 crore. The massive virtual Social event raised Rs 46 million in seven days, demonstrating the country's ability to raise awareness. Virtual entertainment also plays an important role in natural disasters and other disasters by uncovering crimes and improving response.

While online social media gives quiet people a space to express their ideas and work with a study group, there is another side to the story. While his work in subverting the oppressive system and encouraging radicalism has been praised, his potential to incite bad thoughts and behavior has been problematic. Situations where content becomes more transformative than expression undermine freedom of speech and expression. In a country like India, which has different languages, traditions, and strict beliefs, it is a fact that gambling can turn into a big problem over a small incident. Under these conditions, government regulation of online entertainment content has become key to preserving public friendliness. Controlling the free and potentially dangerous nature of online entertainment is still an experiment.

"Freedom of speech and expression is one of the most abused rights in recent times", retired Chief Justice of India, S A Bobde, responded to the Centre's affidavit in the Tablighi case (Rautray 2020), addressing the requirement for care in the face of increased cybercrime and criminal activity through social media. These crimes include insult, invasion of privacy, criminal activity, hate speech and racial discrimination, rioting, torture, theft, harassment, victimization and bullying. One of the media's concerns is to target people and spread rumors to tarnish their public image; is that perpetrators often use freedom of speech and expression to protect themselves. The worst forms of abuse occur in comments on social media pages, causing many people to lose harmony and peace in their lives, becoming the target of faceless people for torture. Problems in society have worsened with the emergence of toxic people who spread fake news, spread negative thoughts, hatred and injustice. Cases of communal violence and mob lynchings caused by social media have highlighted the importance of addressing this issue. More importantly, the investigation revealed that hateful content posted on social media played a role in fueling communal violence, as seen in Muzaffarnagar.

A recent incident in Kerala involved the arrest of two YouTuber brothers for creating chaos in a government office with the support of their followers. Following the arrest of the brothers, two men resisted the attack by speaking out and were placed on file for investigation. These cases underline the need for social equality to avoid undermining the fundamental right to freedom of speech and expression. The challenge is to create a framework that prohibits abusive practices while protecting this fundamental right.

2.6 Challenges of Content Moderation

Achieving balanced content through the entertainment network presents complex challenges arising from the potential, diversity, and value of user-generated content.

2.6.1 Multiple and Fast

The size of user-generated immersive virtual entertainment scene content always represents a good test. Creating a large volume in a really short time requires the success and innovation of the production process. The rapid proliferation of online entertainment content further highlights the need for quick and direct reporting of damaged products or problems.

2.6.2 *Related Subtleties*

The basic rules of management and decision-making history are a good test of good management. Confrontation subtleties are inherent in customer-generated content. Determining right and wrong rhetoric requires a good understanding of the precise context, purpose, social pressures, and quality of social responses. Failure to appreciate content on this topic can lead to distraction or a dangerous impact on the content, highlighting the difficulty of this task. Changing legal and social standards: Virtual entertainment operates worldwide, but legal and social standards vary from region to region. It is an attempt to reconcile different legal standards and find a compromise between various points of view. Levels should explore these changes to manage recycling sustainably and fairly.

2.6.3 *Robotics and Human Discovery*

To manage the large volume of content, the platform frequently employs automated content management systems. However, relying solely on automation may result in errors, glitches, and unintended removal of valid content. It is imperative to strike a balance between automated and human review processes to ensure the accuracy and impartiality of the content (Kemp 2021).

2.7 Ethical Issues

2.7.1 *Abuse of the Laws*

The right to free speech is fundamental, but regulating harmful content requires limits on that freedom. The challenge lies in drawing a line between acceptable speech and harmful speech (hate speech, misinformation, or incitement to violence). And to restrict the abuse of freedom of speech, these laws are necessary. But police or political parties can abuse these laws and regulations to fulfill their own agendas, as was the instance in case of Shreya Singhal.

2.7.2 *Infringement of Right to Privacy*

Excessive monitoring or censorship could infringe on privacy rights and stifle free speech, especially in politically sensitive situations. The line between necessary oversight and surveillance for political control becomes blurred.

2.8 Here Are Some Key Considerations in Balancing Freedom of Expression and Harmful Speech

Legal processes concerning freedom of expression acknowledge that this right has limitations and must be regulated accordingly. Various laws, including civil codes and fundamental freedoms, define these boundaries within which expression can occur. These limitations aim to safeguard public interests, public safety, individual freedoms, and reputations, while also preventing harm. It's essential to carefully analyze evidence and apply these restrictions to strike a balance between freedom of expression and the necessity to address negative consequences.

One of the most difficult issues in maintaining this balance is characterizing and identifying damage products. Vague or ambiguous concepts can lead to inconsistencies and misunderstandings in hedonic regulation. Establishing standards and strict rules forms the basis for clarifying which doctrines are considered unsafe and need mediation. Setting expectations plays an important role in measuring the impact of a message. Details such as the goals behind the speech, its social and relational content, and its intended impact on individuals or minority groups should be carefully analyzed. Negative and influential reviews are important in determining the correct expression of reviews and comments that incite violence, encourage criticism, or target individuals or organizations. The level of virtual entertainment should be managed appropriately, taking into account the main points when evaluating and watching content. It is equally important to respond appropriately to negative feedback. Factors such as the seriousness of the incident, the speaker's intent, and cultural influence should be taken into account when determining appropriate action. Responses to negative comments should demonstrate consistency and solidarity to avoid accusations of injustice or unfair treatment (Sircar 2020).

Finding some compromise between the teaching and management of negative speech in the online entertainment scene is still an ongoing experiment. This requires constant evaluation, refinement of ideas and practices, and recognition that the computing field is in a constant state of change. By combining legal process, facts and evaluation of content, a safe and competent computer will be created.

2.9 Case Laws

Several legal cases have set significant precedents and interpretations concerning social media and freedom of speech in India:

In the Indian National Congress (I) v. Union of India (2017) case, the Supreme Court validated Article 66A but offered clarifications on its interpretation. It determined that online speech could only be limited if it endangered public order or instigated violence. The court underscored the importance of striking a balance between freedom of expression and preserving civil order.

In Kamlesh Vaswani v. Union of India (2013), the court addressed the matter of blocking websites featuring inappropriate content, particularly child pornography. It stated that platforms like social media have a duty to recognize and block access to such content to protect children from harm.

In Faheema Shirin R.K. v. Kerala (2019), the court affirmed that individuals should be able to express themselves freely on social media without fear of censorship or opposition. It stressed the importance of allowing people to voice their opinions without hindrance.

In Maheshwari v. Union of India (2019), a case involving allegations of sharing objectionable content on social media, the court clarified that individuals cannot be held responsible for sharing or reposting content unless there is clear evidence of intent to promote hate speech or cause harm.

2.10 Conclusion

The age of virtual entertainment has revolutionized communication, making the right to freedom of expression a major issue in the computer age. While India recognizes the right to freedom of expression as a fundamental right, it also imposes necessary restrictions to protect public interest and equality. Investigating legal restrictions on freedom of expression in online entertainment requires equal care to ensure that the choice of opinion is protected while preventing the spread of bad language. Regular debates and administrative and judicial reforms are helping to improve India's legal system and establish a good balance (Marlin 2002). Collaboration between the government, service providers and consumers is essential to foster a new and healthy relationship in the country. Service providers, in particular, play an important role in supporting government policies by ensuring the timely disclosure of new information. Despite the changes, social media is still being used for criminal purposes that deserve more regulation.

It is important to strike a fair balance with regulations designed to address emerging issues without affecting fundamental human rights. Realizing the inadequacy of existing cyber laws in India, there is an urgent need to amend Section 66A of the IT Act. In a country as large and diverse as India, there are challenges in establishing guidelines for online entertainment. Programs include seeking input from residents and integrating good governance models from different countries. To achieve this goal, public authorities may form a committee of specialized experts. The group will discuss different aspects of the use of virtual entertainment, explain its positive and negative characteristics, and present the management models necessary to maintain social equality without security.

References

Anuradha Basin v. Union of India (2020 SCC 25) https://globalfreedomofexpression.columbia.edu/cases/bhasin-v-union-of-india/. Accessed 31 Oct 2024

Das M (2017, July 8) Social media posts trigger seven communal riots in a month in West Bengal. The Economic Times. https://economictimes.indiatimes.com/news/politics-and-nation/social-media-posts-trigger-seven-communal-riots-in-a-month-in-west-bengal/articleshow/59496771.cms?from=mdr. Accessed 21 Oct 2024

Faheema Shirin R.K v. State of Kerala and Ors (2019 WP(C) No. 19716 of 2019(L)). https://globalfreedomofexpression.columbia.edu/cases/shirin-r-k-v-state-of-kerala/. Accessed 22 Oct 2024

Fisher M (2012) When is government web censorship justified? An Indian horror story. The Atlantic. http://www.theatlantic.com/international/archive/2012/08/when-is-government-web-censorship-justified-an-indian-horror-story/261396/. Accessed 23 Oct 2024

Government of India (2000) Information Technology Act, 2000 (No. 21). https://www.indiacode.nic.in/handle/123456789/1999. Accessed 7 Dec 2024

Gupta S (2012, November 26) Arrested for Facebook posts, they spent 12 days in jail, lost their Air India jobs. NDTV. https://www.ndtv.com/mumbai-news/arrested-for-facebook-posts-they-spent-12-days-in-jail-lost-their-air-india-jobs-505559. Accesses 7 Nov

Indian National Congress v. Union of India. (2017 WPCR No. 53). https://indiankanoon.org/doc/71183002/. Accessed 10 Oct 2024

Kamath A, Kathpalia V (2020) India takes a tough stand on neighboring apps. The National Law Review. https://www.natlawreview.com/article/india-takes-tough-stand-neighbouring-apps. Accessed 15 Nov 2024

Kamlesh Vaswani v. Union of India (2013 WP (C) No. 177) https://indiankanoon.org/doc/146597737/. Accessed 9 Nov 2024

Kemp S (2021, February 11) Digital 2021: India. Data Reportal. https://datareportal.com/reports/digital-2021-india. Accessed 3 Dec 2024

Kemp S (2023) DIGITAL 2023: India. Data Reportal. https://datareportal.com/reports/digital-2023-india#:~:text=The%20state%20of%20digital%20in%20India%20in%202023&text=There%20were%20692.0%20million%20internet,percent%20of%20the%20total%20population. Accessed 17 Nov 2024

Maheshwari v. Union of India. (2019 Writ No. 1257/2019). from https://indiankanoon.org/doc/65796/. Accessed 29 Oct 2024

Marlin R (2002) Propaganda and the ethics of persuasion. Broadview Press

Parmar R (2012, November 20) 21-year-old girl held for Facebook post questioning Mumbai's Bal Thackeray shutdown. The Times of India. https://m.timesofindia.com/india/21-year-old-girl-held-for-facebook-post-questioning-mumbais-bal-thackeray-shutdown/articleshow/17276979.cms. Accessed 21 Oct 2024

Raj R (2008) Defining contours of press freedom in the backdrop of the National Emergency of 1975. Indian J Integrated Res Law 2(1):155–160

Rautray S (2020, October 8) Tablighi case: freedom of speech and expression most abused right in recent times, says SC. The Economic Times. https://m.economictimes.com/news/politics-and-nation/tablighi-case-freedom-of-speech-and-expression-in-most-abused-right-in-recent-times-says-sc/articleshow/78550618.cms. Accessed 16 Oct 2024

Shreya Singhal v. Union of India (2015 AIR SC 1523). Retrieved 1st November, 2024

Sircar N (2020, March 4) Not all communal riots are local. Social media is now making them national. The Print. https://theprint.in/opinion/not-all-communal-riots-local-social-media-is-making-them-national/375277/. Accessed 28 Oct 2024

United Nations General Assembly (2011) Report of the Special Rapporteur on key trends and challenges to the right of all individuals to seek, receive and impart information and ideas of all kinds through the Internet (A/HRC/17/27). http://daccess-dds-ny.un.org/doc/UNDOC/GEN/G11/132/01/PDF/G1113201.pdf?OpenElement. Accessed 18 Oct 2024

Homa Bansal is Assistant Professor of Law at the Panjab University Regional Centre, Ludhiana, where she has been working since December 2021. She is currently pursuing a Ph.D. in Law from Punjabi University, Patiala, with her research focusing on Fundamental Duties in the Indian Constitution: Comparative Analysis with China and Japan. She holds an LL.M. from the Rajiv Gandhi National University of Law, Patiala, and has qualified for UGC/NET in Law. Her academic background includes a B.A.LL.B. (Honours) from the University Institute of Legal Studies, Panjab University, Chandigarh. Homa has accumulated over four years of teaching experience, having previously worked at Lovely Professional University, Phagwara, and Bajwa Coaching Institute, Patiala, where she taught subjects such as Civil Procedure Code, Indian Evidence Act, Family Law, and Contract Law. She has also held administrative roles, including serving as Editor of the *PURCL Student Law Journal* and working as a student mentor. Her academic contributions include publishing articles on topics such as child rights, live-in relationships, prisoners' rights, and dowry. She has presented papers at numerous national and international conferences on subjects ranging from social media regulation to constitutional law.

Additionally, she has participated in faculty development programmes and workshops designed to enhance her pedagogical skills. Homa is proficient in Microsoft Office, is adaptable to changes, and has strong interpersonal skills. She is fluent in English, Hindi, and Punjabi and is married.

Jesus Goyal is an accomplished individual with a strong background in leadership, academics, cultural activities, and event management. He has held several key leadership positions, including Vice Head Boy at DPS Bathinda, Discipline Secretary, and Convener of the Moot Court Society at the Panjab University Regional Centre. Additionally, he has actively participated in the National Service Scheme (NSS) for three years, coordinating multiple events, including Youth Fests and Oath Ceremonies. His legal achievements include winning prestigious national moot court competitions, such as the Panjab University SSG National Moot Court Competition (first position and Best Memorial) and the Justice R.R. Aggarwal National Moot Court Competition (first position). He has also participated in international moot court competitions, including the 13th Amity International Moot Court Competition and the SGTU International Moot Court Competition. Beyond competing, he has played a vital role in organizing legal events, notably as Co-Convener of the 4th PURC National Moot Court Competition in 2024. Jesus is a certified mediator by the International Mediation Institute and has served as Jury Member at the Solaris SDG Summit Moot Court Competition. He has chaired panel discussions on significant legal topics such as remission policies and constitutional rights. His involvement in academia extends to organizing and facilitating client counselling competitions, declamation contests, and legal panel discussions at law fests and seminars. Beyond the legal sphere, Jesus has been actively engaged in cultural activities, participating in dance competitions, fashion shows, and Model United Nations (MUN) conferences. His MUN experience spans roles as a delegate, organizer, and Executive Board Member, where he has led marketing and finance teams and received multiple accolades. An accomplished athlete, Jesus has represented his state in badminton and football. His commitment to social responsibility is evident through his participation in cleanliness drives, plantation initiatives, drug abuse awareness programmes, and blood donation camps. He has also made significant contributions to event management, overseeing TEDx events, guest lectures, webinars, and moot court competitions. With his extensive leadership experience, academic excellence, and dedication to social initiatives, Jesus Goyal continues to make a meaningful impact across multiple domains.

Chapter 3
Data Security and Consumer Trust: A Legal Examination of Online Consumer Protection Regulations in the Indian Context

Shalinee Vishwakarma and Karan Goyal

Abstract The exponential growth of e-commerce in India has transformed the consumer landscape, necessitating a robust legal and regulatory framework to safeguard the rights and welfare of online consumers. This article aims to protect the rights and interests of consumers who make purchases online. This research paper examines the evolving online consumer protection regulations in India and the best practices adopted worldwide. It examines the legislative and regulatory measures introduced to address the unique challenges posed by the digital marketplace. The study examines key laws, including the Consumer Protection Act of 2019 and the Information Technology Act of 2000. It examines whether these laws are effective in ensuring online shopping is safe and trustworthy.

Keywords Data security · Personal data · Data protection · Consumer data · E-commerce · Consumer protection

---

The chapters in this books are presented in the "International Conference on Legal and Economic Synergy for an Equitable Society, organised by the University Institute of Laws at Panjab University Regional Centre, Ludhiana, India on February 1st, 2024".

S. Vishwakarma (✉)
Jindal Global Law School, OP Jindal Global University, Sonipat, Haryana, India
e-mail: shalineevishwakarma2907@gmail.com

K. Goyal
Panjab University, Chandigarh, India
e-mail: goyakkaran96@gmail.com

A. Virk et al. (eds.), *Law, Economy, and Society: Insights into Legal and Economic Issues in India*, Accounting, Finance, Sustainability, Governance & Fraud: Theory and Application, https://doi.org/10.1007/978-981-95-3361-9_3

3.1 Introduction

The proliferation of Information and Communication Technology has significantly impacted several facets of human existence, including the trade of commodities and services. The emergence of information technology has led to the development of revolutionary technologies, including electronic data exchange and cash transfer. These technologies, facilitated by the Internet, have evolved into commercial activities known as e-commerce, which Information Technology enables. E-commerce involves conducting business operations electronically, especially using technologies such as EDI (Electronic Data Interchange) and EFT (Electronic Funds Transfer), within the internet domain. E-commerce refers to the exchange of goods or services over electronic systems, including the Internet and other computer networks. Electronic commerce utilizes various technologies, including electronic money transfer, supply chain management, Internet marketing, online transaction processing, electronic data interchange (EDI), inventory management systems, and automated data collection systems. Contemporary electronic commerce often involves using the World Wide Web during the various stages of a transaction. However, it can also utilize other technologies, such as email, mobile devices, and telephones (Khare and Rajvanshi 2013).

Although e-commerce transactions present several obstacles, ensuring consumer safety in these transactions is of utmost significance. Insufficient safeguards for consumers can have negative effects on fair competition, hinder the unrestricted dissemination of accurate information in the market, and provide opportunities for consumer fraud and deceit. Hence, it is very important to look at and study how people are kept safe from bad things, and e-commerce, mainly from a legal perspective. Keeping people safe, in a big way, aims to stop people from many unfair trade practices. The reason for this safety, as told in the case of Raghubir Singh v. Thakurain Sukhraj Kuar (AIR 1939), is to stop people from being taken advantage of by the business world and stop many bad business practices. Business groups, being very organized, knowing more, and having power in the market, have ways to easily take advantage of people. So keeping people safe becomes necessary to protect the ones hurt most from these bad ways.

3.2 Importance of Ensuring Online Consumer Protection

Consumer protection is a significant effort. It safeguards people from unfair business practices. It stops companies from exploiting or harming people's rights in competitive markets. We must note that businesses have a big advantage over consumers in commercial dealings. They have more expertise, knowledge, and market power. This advantageous position can lead to consumer exploitation (Khare and Rajvanshi 2013). This is why consumer protection is so necessary. Several reasons contribute to the crucial need for consumer protection, including:

- Fraud and Scams: Online platforms are vulnerable to many fraudulent activities. This includes phishing, online scams, and fake websites. These pose threats to consumers. They may suffer financial losses or have personal information compromised.
- Data Privacy Concerns: Internet firms may improperly use personal data. This creates big privacy problems. Customers often don't know how their data is used. Data breaches and identity theft can happen due to a lack of protection.
- Counterfeit Products: E-commerce platforms can be places for selling counterfeit or substandard products. Consumers may receive items that greatly differ from advertised quality standards.
- Misleading Advertisements: Online advertising may be deceptive. It often has inflated claims about items or services. Consumers relying on misleading information may make purchase choices leading to unhappiness.
- Difficulties in Dispute Resolution: Resolving disputes with online sellers or service providers proves challenging due to the lack of effective mechanisms and standardized processes for grievance redressal, leaving consumers frustrated.
- Consumer Education Gap: Many consumers lack awareness about their rights and responsibilities in the online marketplace, making them more susceptible to exploitation. Knowledge gaps regarding online consumer protection laws and best practices exacerbate this vulnerability.
- Complexity in Refund Processes: The process of obtaining refunds for defective or unsatisfactory items can be complicated, since some online platforms have extensive refund policies or fail to immediately resolve refund requests.
- Inadequate Regulation of Marketplaces: The regulatory structure overseeing e-commerce may not keep pace with the fast-changing nature of online transactions, leading to deficiencies in safeguarding consumer rights and difficulties in holding online platforms responsible for unethical behaviour.
- Delivery Issues: Consumers often encounter problems like delayed deliveries, non-delivery of products, or receipt of damaged goods. Inconsistent delivery standards and logistical challenges contribute to these issues.
- Fake Reviews and Ratings: Manipulation of product reviews and ratings is a prevalent issue on online platforms. Fake reviews can mislead consumers and impact their purchasing decisions, as many rely on these reviews for product information.

Understanding the legal basis for consumer protection is essential to evaluating the effectiveness of consumer protection regulations. This means thoroughly examining the constitutional and legislative laws that establish the foundation of online consumer protection in India. The goal is to clarify the rights and safeguards available to consumers (Finserv 2022).

3.3 Evolution of Online Consumer Protection in India

Amidst a digital world where digital transactions and virtual interactions are prevalent, ensuring the security of data and building customer confidence have become crucial. As technology has advanced, the difficulties associated with protecting consumers' personal information online have also changed. This investigation examines the legislative landscape of online consumer protection rules in India, closely examining the steps used to safeguard data and foster customer confidence.

The evolution of online consumer protection in India has witnessed significant changes throughout time. Starting with basic principles, the legal system has adjusted to address the complex issues presented by the digital era (Adgulwar 2021). This section provides a comprehensive overview of the important milestones in the development of online consumer protection rules in India, highlighting several governmental efforts in this area.

- **Information Technology Act, 2000:** (Government of India 2020) The Information Technology Act of 2000 stands as an early legislative initiative in India, addressing concerns related to electronic commerce and online transactions. While its primary focus rested on electronic records and digital signatures, it laid the groundwork for tackling electronic fraud and concerns related to online consumer protection.
- **Consumer Protection Act, 2019:** Supplanting the erstwhile Consumer Protection Act of 1986, the Consumer Protection Act of 2019 ushered in provisions specifically tailored for e-commerce transactions and online consumer rights. This legislation birthed the Central Consumer Protection Authority (CCPA), tasked with the promotion, protection, and enforcement of consumer rights.(Press Information Bureau Government of India 2020)
- **National Consumer Helpline (NCH):** The National Consumer Helpline (NCH), initiated by the Department of Consumer Affairs, offers a platform where consumers can file complaints and seek information regarding their rights. It plays a crucial role in resolving consumer grievances related to online transactions.
- **E-commerce Guidelines:** In 2020, the Ministry of Consumer Affairs issued specific guidelines for e-commerce businesses. These rules cover aspects like displaying product details accurately, transparent pricing, and procedures to address complaints. Adhering to these principles is mandatory for e-commerce companies to ensure consumer protection.
- **Digital India Initiative:** The Digital India effort, driven by the government, aims to transform India into a technologically advanced and empowered nation. This effort emphasises an increased emphasis on strengthening measures to safeguard online consumers, including promoting digital literacy and conducting awareness campaigns to prevent online fraud.
- **Competition Commission of India (CCI):** The Competition Commission of India (CCI) primarily aims to promote fair competition and also plays a role in protecting consumer interests. The CCI conducts thorough investigations into anti-competitive behaviours, specifically focusing on the e-commerce industry,

thereby indirectly enhancing online consumer protection. There have been several instances of internet data breaches affecting customers.

3.4 Case Studies on Data Breach

Data breaches in e-commerce pertain to the illicit entry, revelation, or procurement of sensitive data associated with users or consumers from online retail platforms. The repercussions of these breaches may be highly detrimental for both customers and the impacted organisations (Chawla and Kumar 2022). Multiple categories of data are susceptible to compromise, such as personal information encompassing names, addresses, email addresses, phone numbers, and other personally identifiable information (PII). Financial information, including credit card details, bank account information, and other financial data, is frequently targeted. Additionally, login credentials are also sought after. Unauthorised individuals may get access to accounts by compromising usernames and passwords (Justia 2023).

Data breaches occur due to a multitude of underlying factors. Cyberattacks can occur when hackers employ techniques such as phishing, malware, or exploiting software vulnerabilities to gain unauthorised access. Breaches in organisations can result from actions taken by employees, contractors, or individuals with insider access, as well as from vulnerabilities in third-party systems. Data breaches can also occur as a result of vulnerabilities in the security systems of third-party suppliers or service providers.

These breaches can result in several outcomes for customers, including money losses, identity theft, or privacy problems. Personal data breaches can give rise to privacy issues for people. Stolen financial information can result in unauthorised transactions and financial consequences for consumers.

3.5 Instances of Data Breaches in India

India has had many instances of online data breaches, each highlighting distinct difficulties and emphasising the crucial requirement for strong cybersecurity measures and legal protections. Here are some notable examples:

- **Aadhaar Data Breach (2023):** (Livemint 2023) In 2023, the Aadhaar database of the Unique Identification Authority of India (UIDAI) had a massive data breach. The security compromise resulted in the unauthorised disclosure of the personal information of more than one billion Indian people, including their names, Aadhaar numbers, and demographic data. The illicit acquisition of Aadhaar data in exchange for payment has sparked significant apprehension over the security and confidentiality of highly sensitive personal information within the largest biometric identification system in the world. The occurrence triggered swift remedial measures by the UIDAI, igniting a reassessment of legislation pertaining to the safeguarding of data.

- **Zomato Data Breach (2017):** In 2017, Zomato (Thomas 2017), a famous food delivery company, had a data breach. This breach affected many customers. Their email addresses and encrypted passwords got leaked. This event showed problems in Zomato's security system. Zomato quickly told customers about the breach. They stressed the importance of open and honest communication after the attack.
- **Reliance Jio Data Breach (2017)** (Business Today 2017)**:** Reliance Jio, a big telecom company in India, had a data breach in 2017. People got access to customer data without permission. Names, mobile numbers, and email addresses got published. This showed how important it is to secure telecom systems to protect customer information.
- **Air India Data Breach (2021):** In 2021, Air India, India's biggest airline, had a huge data breach. Many customers were affected. The hack entailed illicit entry into travellers' personal data, encompassing passport particulars and credit card details. This occurrence prompted apprehensions over identity theft and financial crime, igniting deliberations concerning organisational culpability and adherence to data protection standards (Mehta and Kumar 2021).
- **State Bank of India (SBI) Yono App Data Leak (2019):** In 2019, the State Bank of India's Yono app, which is a digital banking platform, had a data breach that resulted in the exposure of customers' personal information. The compromise entailed the unintentional disclosure of source code and unauthorised access to a testing environment. SBI swiftly addressed the situation, reassuring consumers that no critical financial information was exposed. They emphasised the need for safe software development procedures (Whittaker 2019).
- **Dr. Reddy's Laboratories Cyber Attack (2020):** (Livemint 2020) A prominent pharmaceutical firm in India had a cyber-attack in 2020. This incident caused disruptions to their operations and prompted worries over the security of healthcare data. The occurrence triggered inquiries into causes and ramifications for the pharmaceutical business, highlighting the susceptibility of crucial industries to cyber assaults.

These case studies demonstrate the varied types of data breaches in India, underscoring the need for strong cybersecurity, prompt incident response, and ongoing efforts to enhance data protection legislation and enforcement. Every data breach helps discuss data security, privacy, and companies' legal duties to protect private information.

3.6 Data Security Frameworks in India

Online consumer protection is fundamentally dependent on data security. India has established many frameworks and policies to guarantee the secure management of customer data by corporations. This section examines the current state of data protection, specifically focusing on the Personal Data Protection Bill and other pertinent rules. It analyses the potential effects of these restrictions on internet businesses and consumers.

India has been proactively designing and deploying data security standards to tackle the increasing difficulties in the digital environment. The latest Personal Data Protection Act 2023 (Chawla and Kumar 2022) is an important law. It governs how people handle personal data in the country. The law gives rules for keeping data safe. It protects the rights of people. It says what businesses must do with the personal data they have. It also says there will be a Data Protection Authority (DPA).

The DPA will make sure people follow data protection laws. It will deal with data breaches too. The Information Technology (Reasonable Security Practices and Procedures and Sensitive Personal Data or Information) Rules, 2011, are part of the Information Technology Act, 2000. These rules give guidelines for good security practices and procedures. They are specifically for managing sensitive personal data or information.

The rules say what organizations must do when they collect and handle this info. They must have security protocols in place. The National Cyber Security Policy came out in 2013. It tries to make a safe digital environment. It wants to make the country's cybersecurity stronger. It focuses on making vital information infrastructure more secure. It promotes the use of good cybersecurity practices too. The Reserve Bank of India (RBI) often gives advice and policies.

They aim to improve cybersecurity in the financial industry. The rules cover making online transactions safer. They try to reduce fraud. They protect the financial system from cyber-attacks. In India, many organizations use the ISO/IEC 27001 standard. This standard manages information security systems. It provides a systematic way to create and maintain good information security management. The Indian Computer Emergency Response Team (CERT-In) is the main agency for cybersecurity issues. CERT-In gives recommendations, warnings, and advisories. It helps organizations and the public be more aware of cybersecurity. It helps them respond better, too. The Ministry of Health and Family Welfare makes rules for the health sector. There is a Health Data Management Policy. Its purpose is to ensure health data is managed safely.

This is important as digital health projects grow. The proposed Data Empowerment and Protection Architecture (DEPA) focuses on giving people control over their personal data. It has consent frameworks, data protection, and user control. Although India has made great progress in data security, the law keeps changing. Organizations should stay updated on the latest developments. They must follow new standards in data security, too.

3.7 Laws in India on Dispute resolution for E-Commerce

Laws in India help solve issues related to e-commerce. They target scams and data breaches efficiently. The Consumer Protection Act 2019 gives a legal way to handle customer complaints about e-commerce scams or data breaches. Customers can file complaints at consumer forums based on the cost of the product or service.

The National Consumer Helpline (NCH) allows people to lodge complaints and get advice on e-commerce scams. Complaints can be filed online or through a toll-free number. The Ministry of Consumer Affairs has rules for e-commerce platforms to handle complaints fairly and transparently. The Reserve Bank of India (RBI) oversees digital transactions and electronic banking. Customers facing unauthorized transactions or online fraud can complain to the RBI ombudsman.

The proposed Data Protection Authority (DPA) will manage data breach complaints once the Personal Data Protection Bill 2019 is implemented. Officials appointed under the IT Act 2000 can resolve disputes related to electronic transactions like data breaches. They can impose fines for violations. Consumers can seek help from various consumer courts and ombudsman schemes, like the Banking Ombudsman for banking issues.

Arbitration and mediation provide alternative ways to resolve e-commerce disputes (Juris Centre 2023) outside courts. Many e-commerce platforms have their own dispute resolution systems. In case of illegal activities, consumers can file complaints with the local police. Consumers should know their rights, keep transaction records, and first try resolving issues directly with the e-commerce platform. If unsuccessful, they should approach the appropriate regulatory body or consumer forum. Staying updated on changing laws and regulations is crucial (Coyle 2022).

3.8 Comparison of USA, EU and Chinese Laws

Comparison of USA, EU, and Chinese Laws. In the present era of an interdependent global economy, safeguarding consumer data has emerged as a crucial priority for both governments and corporations. With the progression of digital technology, there is an increasing requirement for strong legislation to protect individuals' sensitive information from unauthorized access and exploitation. This comparative research aims to examine the consumer data protection systems in the United States, the European Union, and China.

- **United States:** The United States has a sectoral approach to data protection, where different laws regulate distinct sectors and types of data. The Gramm-Leach-Bliley Act (GLBA) is a key federal policy that specifically targets consumer data protection in the banking industry. GLBA mandates that financial institutions must establish measures to preserve client information and uphold consumer privacy. The Health Insurance Portability and Accountability Act (HIPAA) is an important legislation in the United States that protects health-related information. HIPAA mandates stringent regulations on healthcare providers, insurers, and their business affiliates to guarantee the privacy and protection of individuals' health information. Nevertheless, the lack of a comprehensive federal data protection legislation has resulted in a fragmented set of rules. State-level legislation, such as the California Consumer Privacy Act (CCPA) and Virginia's Consumer Data Protection Act (CDPA) (State of California - Department of Justice 2023), has been introduced to address the existing deficiencies. These state statutes confer upon consumers the entitlement to obtain, erase, and decline participation in the transaction of their personal data.

- **European Union:** The European Union has implemented the General Data Protection Regulation (GDPR) (General Data Protection Regulation (GDPR) 2022) to provide a comprehensive and standardized strategy to safeguarding consumer data. The General Data Protection Regulation (GDPR), which came into effect in 2018, is applicable to all member states of the European Union (EU) and serves to harmonies data protection legislation throughout the EU. With the implementation of GDPR, individuals are granted enhanced authority over their personal data. Individuals possess the entitlement to be informed about the way personal data is handled, the entitlement to retrieve it, and the entitlement to have it deleted. Companies must include privacy by design and default, which entails integrating data protection into their systems and procedures from the beginning. The General Data Protection Regulation (GDPR) requires some organizations to designate Data Protection Officers (DPOs), who are responsible for overseeing internal data protection processes. The legislation enforces strict penalties for failure to comply, including fines that can amount to 4% of a company's worldwide annual revenue.
- **China:** China has adopted a government-centric approach to data protection, which differs from the approaches used by the EU and the U.S. The implementation of the Personal Information Protection Law (PIPL) in 2021 (China Briefing News 2021) represents a significant advancement in the regulation of personal information collection and processing in China. PIPL incorporates concepts that are similar to those found in GDPR, such as the necessity of obtaining user consent for data processing and granting users the right to view and rectify their personal information. Additionally, it contains restrictions for the movement of data across borders, which require evaluations and authorization from authorities. In addition, China has enacted the Cybersecurity Law to govern the administration of network products and services. This legislation mandates that operators of essential information infrastructure must retain personal information and crucial data inside the borders of the nation. Additionally, it grants the government the authority to carry out security evaluations on these businesses.
- **Comparative Analysis:** The three areas demonstrate different methods of safeguarding customer data. The United States implements a sectoral model that focuses on various industries but does not have a complete federal structure. On the other hand, the European Union's General Data Protection Regulation (GDPR) offers a comprehensive and strong set of rules that are applicable to all member states. It focuses on protecting individual rights and imposes significant

penalties for failure to comply. China's approach is characterized by government monitoring, as seen by the implementation of the Personal Information Protection Law (PIPL) and the Cybersecurity Law, which provide authorities with substantial control over the movement and handling of data. The focus on localizing data within China is in line with the government's wider plan to strengthen cybersecurity and ensure national control over data. Ultimately, as the digital environment progresses, the methods of safeguarding customer data also advance. While the United States deals with a fragmented set of regulations specific to different sectors, the European Union develops a robust and consistent framework, while China prioritizes government control. Each method mirrors the distinct legal, cultural, and political circumstances in which these rules are developed. Continual assessment and modification of these frameworks will be essential in guaranteeing the efficient safeguarding of consumer data worldwide, as technology progresses.

3.9 Way Forward

To effectively implement legislation that safeguards online consumers, a comprehensive strategy is required to tackle the ever-changing difficulties of the digital environment. In order to bolster online consumer protection, there are various important recommendations that may be taken into account. Primarily, it is essential to implement and enhance comprehensive regulations for data protection. This legislation shall unambiguously delineate consumer entitlements for the gathering, handling, and retention of personal information, encompassing provisions for well-informed agreement, data reduction, restriction of usage, and the entitlement to have one's data erased.

It is equally crucial to require companies to swiftly inform both customers and appropriate regulatory agencies in the case of a data breach. This guarantees transparency, enabling users to proactively take precautionary actions. Implementing more stringent cybersecurity protocols for firms that handle consumer data, including encryption, access limits, frequent security audits, and proactive steps against emerging threats, is crucial in order to protect sensitive information.

In order to give consumers more power, it is vital to implement strong instructional programmes. These programmes should prioritise increasing knowledge about online hazards, privacy configurations, and entitlements. This may be achieved by efforts that educate individuals on identifying phishing attempts, comprehending privacy policies, and making well-informed decisions when using the internet. In addition, it is imperative to enforce explicit and comprehensible terms of service and privacy policies to educate consumers on the utilisation and dissemination of their data (Satyan 2015). It is essential to enhance and simplify the systems for addressing grievances, making sure that customers have convenient and effective ways to resolve disputes, lodge complaints, and obtain compensation for damages resulting from data breaches or unfair corporate practices. Robust regulatory supervision, which encompasses frequent audits and investigations into the manner in which firms manage data, is crucial for ensuring adherence to regulations.

To tackle the difficulties originating from the transfer of data across borders, it is necessary to develop collaboration in the field of cross-border data protection. Promoting cooperation among international regulatory authorities can effectively address global challenges pertaining to the protection of online consumers. Stringent sanctions should be implemented to discourage organisations from neglecting data security and failing to comply with data protection requirements.

It is necessary to continually update regulations in order to stay up to speed with technology improvements and new dangers. It is necessary to establish mechanisms that enable quick and flexible responses to emerging difficulties, ensuring that the regulatory framework can adapt to changes in the digital ecosystem. Additional ways to increase online consumer safety include promoting the adoption of privacy by design principles among firms, incorporating provisions for whistle-blower protection, fostering collaboration between the public and private sectors, guaranteeing compliance with accessibility requirements, and formulating legislation that is technologically neutral.

Executing these recommendations necessitates a cooperative endeavour including legislators, regulatory entities, enterprises, consumer advocacy organisations, and technological specialists. Frequent evaluations and revisions of legislation will be crucial in order to anticipate and adapt to the constantly evolving digital economy, establishing a fair and secure environment for online transactions.

3.10 Discussion

The legal review of online consumer protection regulations in India presents a complex scenario with ever-changing difficulties and possibilities. Maintaining a balance between data security and consumer trust necessitates constant adjustment of regulatory frameworks to keep up with the dynamic nature of the digital world. By comprehending the historical developments, legal foundations, and current challenges, stakeholders can collaborate to cultivate a secure and trustworthy online environment for consumers in India. This involves implementing robust measures to safeguard personal information and prevent unauthorized access or misuse. Furthermore, transparent policies and clear communication channels empower consumers to make informed decisions and raise concerns effectively. Simultaneously, fostering technological innovation and promoting fair competition contributes to a thriving digital marketplace. A comprehensive approach prioritizes consumer rights, privacy, and security, while facilitating economic growth and digital transformation is crucial.

References

Adgulwar R (2021) Consumer protection and e-commerce in India. Palarch's J Archaeol Egypt/Egyptol 18(10):990–997. ISSN 1567-214X

AIR 1939 Oudh 96 at 99

Business Today (2017, July 13) Reliance Jio files FIR for data breach. https://www.businesstoday.in/industry/telecom/story/reliance-jio-files-fir-for-data-breach-86109-2017-07-13. Accessed Mar 2024

Chawla N, Kumar B (2022) E-commerce and consumer protection in India: the emerging trend. J Bus Ethics 180:581–604. https://doi.org/10.1007/s10551-021-04884-3

China Briefing News (2021, December 29) The PRC Personal Information Protection Law (final): a full translation. https://www.china-briefing.com/news/the-prc-personal-information-protection-law-final-a-full-translation/. Accessed 15 Mar 2024

Coyle M (2022, July 18) What are the disadvantages of ADR?. Lawdit Solicitors. https://lawdit.co.uk/readingroom/adr-disadvantages. Accessed 02 Jan 2024

Finserv B (2022, May 25) E-commerce frauds in India: Types of frauds in e-commerce. https://www.bajajfinserv.in/e-commerce-frauds-to-watch-out-for. Accessed 15 Mar 2024

General Data Protection Regulation (GDPR) (2022, September 27) Official legal text. https://gdpr-info.eu/. Accessed 02 Jan 2024

Government of India (2020) The Information Technology Act, 2000 - India code. https://www.indiacode.nic.in/bitstream/123456789/13116/1/it_act_2000_updated.pdf.Accessed 02 Jan 2024

Justia (2023, September 21) Data breaches & consumers' legal rights to privacy. https://www.justia.com/consumer/identity-theft/data-breaches-privacy/. Accessed 15 Mar 2024

Juris Centre (2023, August 22) Alternative dispute resolution: types, advantages and disadvantages. https://juriscentre.com/2023/08/22/alternative-dispute-resolution-types-advantages-anddisadvantages. Accessed 15 Mar 2024

Khare, Rajiv Professor of Law and Rajvanshi, Gargi Assistant Professor of Law (2013) E-Commerce and Consumer Protection: A Critical Analysis of Legal Regulations. International Journal on Consumer Law and Practice: Vol. 1, Article 6. Available at: https://repository.nls.ac.in/ijclp/vol1/iss1/6. Accessed 15 Mar 2024

Livemint (2020) Dr. Reddy's provides update on cyber-attack. https://www.livemint.com/companies/news/dr-reddy-s-provides-update-on-cyber-attack-11603873684385.html, October 28, 2020. Accessed 15 Mar 2024

Mehta T, Kumar A (2021) Air India says February's data breach affected 4.5 million passengers. Reuters. https://www.reuters.com/world/india/air-india-says-februarys-data-breach-affected-45-mln-passengers-2021-05-21/. Accessed 15 Mar 2024

Press Information Bureau Government of India, Ministry of Consumer Affairs, Food & Public Distribution (2020) For Consumer Protection (E-Commerce) Rules 2020. https://pib.gov.in/Pressreleaseshare.aspx?PRID=1656161. Accessed 15 Mar 2024

Satyan K (2015, July 2) E-commerce and consumer rights: applicability of consumer protection laws in online transactions in India. Soc Sci Res Network (SSRN). https://ssrn.com/abstract=2626027

State of California - Department of Justice (2023, November 9) California Consumer Privacy Act (CCPA). https://oag.ca.gov/privacy/ccpa. Accessed 15 Mar 2024

Thomas A (2017) The Economic Times. (n.d.). Zomato hacked: Security breach results in 17 million user data stolen. https://economictimes.indiatimes.com/small-biz/security-tech/security/zomato-hacked-security-breach-results-in-17-million-user-data-stolen/articleshow/58729251.cms?from=mdr, May 19, 2017. Accessed 15 Mar 2024

Whittaker Z (2019) India's largest bank SBI, leaked account data on millions of customers. TechCrunch, January 30th, 2019. https://techcrunch.com/2019/01/30/state-bank-india-data-leak/. Accessed 15 Mar 2024

Shalinee Vishwakarma holds a BLS LL.B. (Hons.) degree from Dr. DY Patil College of Law, Mumbai University, and an LL.M. in Corporate and Financial Law and Policy from O.P. Jindal Global University. With extensive experience gained through her work at various law firms and with senior advocates, she has developed expertise in Company Law, the Negotiable Instruments Act, the Consumer Protection Act, and Labour Law. She has also assisted in drafting legal opinions, notices, plaints, written statements, and applications. Shalinee is committed to teaching and believes it requires passion and dedication beyond mere professional obligations. Her pedagogical approach emphasizes innovative learning and research-based methodologies. She believes in creating an open and interactive learning environment where students can ask questions and share their curiosity. In her research, Shalinee has focused on several areas of interest, including international commercial arbitration, Indian arbitration, negotiation and mediation, and LGBT rights.

Karan Goyal obtained a B.Com.LL.B. (Hons.) degree from Panjab University, Chandigarh, in 2019. He completed his Master's in Law (LL.M.) in Access to Justice with a special focus on the rights of marginalized sections of society at the Tata Institute of Social Sciences (TISS), Mumbai, in 2020. He is pursuing his doctoral research at Panjab University, Chandigarh, on access to justice vis-à-vis denotified tribes in India. His LL.M. dissertation on "Interface between Adivasi and Criminal Courts" primarily focused on the most marginalized section of Indian society, i.e. tribal people. He has interned with the Human Rights Commission and the Legal Service Authority, and he is enrolled as an advocate with the Bar Council of Punjab and Haryana. His research interests include the criminal justice system and the rights of marginalized sections.

Chapter 4
The Market for 'Disinformation': Investigating the Economic Costs and Motivations of 'Fake News'

B. Karthik Shiva

The truth will always have a market.

Jean Shepard

Abstract Disinformation and the problem of 'fake news' are often deliberated and dissected as a problem from the prism of civil liberties, public safety, platform governance, and technological regulation. The study of 'disinformation' has often focused excessively on its political and social implications but not adequately on the economic cost and motivations of 'fake news', which is frequently underemphasised. There is a cause-and-effect relationship between market economies, the dissemination of 'fake news', and its economic considerations. Social media platforms rely upon paid advertisements as the major source of revenue, which in many cases entails overlooking problematic content, including disinformation and misinformation. Any attempt to address, deal with, and remedy the menace of disinformation without unpacking the 'market' system and economic aspects in this equation, encompassing social media platforms, digital content creators and influencers, netizens, and the government, simply amounts to missing the woods for the trees. Parallelly, a branch of 'disinformation' has emerged that targets companies, brands, and markets, leading to unprecedented economic loss with far-reaching implications. Technologies such as 'deepfakes', machine learning, generative AI, etc., con-

The chapters in this books are presented in the "International Conference on Legal and Economic Synergy for an Equitable Society, organised by the University Institute of Laws at Panjab University Regional Centre, Ludhiana, India on February 1st, 2024".
This chapter is related to author's PhD research thesis, which is yet to be submitted.

B. Karthik Shiva (✉)
The Tamil Nadu Dr Ambedkar Law University, Chennai, India

National Law Institute University, Bhopal, India
e-mail: karthikshiva.phd@nliu.ac.in

A. Virk et al. (eds.), *Law, Economy, and Society: Insights into Legal and Economic Issues in India*, Accounting, Finance, Sustainability, Governance & Fraud: Theory and Application, https://doi.org/10.1007/978-981-95-3361-9_4

tribute to the expansion of financial motivations and costs of 'fake news', opening myriad prospects of productive and disruptive uses. In this broad background, the present paper aims to study and investigate the market for disinformation to truly uncover the economic costs and motivations of 'fake news' in an attempt to regulate them effectively.

Keywords Disinformation · Markets · Deepfakes · Fake news · Social media

4.1 Introduction

Truth is a hallmark of a vibrant democracy. It has both a cause-and-effect relationship in a democracy. It is the lifeblood of democratic frameworks and constitutional systems of governance. The concept of truth has an ethical dimension and an epistemological dimension. In the parlance of morality and ethics, it is considered the opposite of lies; epistemologically, it is the opposite of mistaken beliefs and erroneous information (Rosenfield 2019). In both aspects, the conception of 'truth' forms a fundamental and indivisible aspect of the Rule of law, democratic values, and good governance. The objective idea of what 'truth' means in popular culture, political discourse, and public understanding has witnessed a cataclysmic change in the recent past. The usage of the phrase 'fake news' in an emblematic manner, cutting across political, ideological, and economic spectrums with different understandings, varied meanings and interpretations, often in paradoxical senses, is a reflection of this transition.

Disinformation and the problem of 'fake news' are often deliberated and dissected as a problem from the prism of civil liberties, public safety, platform governance, and technological regulation. The study of 'disinformation' has often focused excessively on its political and social implications but not adequately on the economic cost and motivations of 'fake news', which is frequently underemphasised. There is a cause-and-effect relationship between market economies, the dissemination of 'fake news', and its economic considerations. Social media platforms rely upon paid advertisements as the major source of revenue, which in many cases entails overlooking problematic content, including disinformation and misinformation.

Any attempt to address, deal with, and remedy the menace of disinformation without unpacking the 'market' system and economic aspects in this equation, encompassing social media platforms, digital content creators and influencers, netizens, and the government, simply amounts to missing the woods for the trees. Parallelly, a branch of 'disinformation' has emerged that targets companies, brands, and markets, leading to unprecedented economic loss with far-reaching implications. Technologies such as 'deepfakes', machine learning, generative AI, etc., contribute to the expansion of financial motivations and costs of 'fake news', opening myriad prospects of productive and disruptive uses. In this broad background, the

Fig. 4.1 Spectrums in the information disorder

present paper aims to study and investigate the market for disinformation to truly uncover the economic costs and motivations of 'fake news' in an attempt to regulate them effectively.

The distinction between 'disinformation' and 'misinformation', often commonly used interchangeably, is argued to have different meanings despite an element of cohesion as both refer to 'false information'. However, the degree of divergence pertains to whether the information is spread or shared without intention/awareness regarding its falsity and without intent to cause harm, in which case it is termed' misinformation'. Disinformation, on the other hand, refers to untrue or false information that is circulated or shared with knowledge and awareness of its falsity with the intention to cause or bring about harm through such dissemination.

Thus, disinformation refers to content that is circulated and known to be untrue or not factual, or false information. Misinformation, on the other hand, refers to content that is believed to be true or shared without the intent of misleading, but is found to be untrue or not factual. The key distinction is the intent of the parties with respect to the truth of the matter. Additionally, scholars also propose a third category in the form of 'malinformation', in which truthful information is spread with the intention of causing harm to specific persons. All three spectrums are broadly placed under the omnibus terminology referred to as 'Information Disorder' (Wardle and Derakhshan 2018) (Fig. 4.1).

The present chapter is divided into three parts.

Part I deals with modalities of the disinformation economy and aims to explain and unpack it through the three facets of its costs, price, and profits.

Part II attempts to dissect disinformation through the four major classifications, viz.: Electoral Disinformation, Financial Disinformation, Scientific Disinformation, and Medical Disinformation.

Part III deals with the regulation of the market for disinformation, factoring in the economic forces, motivations, and actors at play so as to ensure that democratic regimes are able to manage and tide over the current 'credibility' crisis.

4.2 Part 1: Unpacking the Disinformation Economy

The market for disinformation works on the general principles surrounding the economics of the market and revolves around the quintessential twin driving forces of 'demand' and 'supply' that serve as its basis (Shumaker 2019). There is a need to focus on these dual aspects in the attempt to truly curb, check, and regulate disinformation and misinformation. Until recently, the core focus has been on the 'demand' side of the problem, with ongoing initiatives that seek to educate internet users and citizens from being misled by 'disinformation' and 'misinformation'. Content flagging, regulation, and moderation, which primarily focus on the 'supply' side of disinformation, have been gaining traction in the recent past. It may take the form of content takedown, adding labels indicating the true context or false nature of statements made and restricting access to repeat offenders who spread 'disinformation' and 'misinformation'. While there can be no serious challenges on social media platforms seeking to educate their users to be wary of 'fake news', the attempt to regulate content that involves disinformation and misinformation entails treading on a tightrope and is fraught with logistical, economic, and legal challenges.

One of the primary reasons why content moderation and regulation is viewed with great circumspection both by social media platforms and internet users is the propensity it poses to governments, social media platforms, and other entities to scuttle free speech. The core argument against content regulation and moderation, with the exception of extreme cases such as hate speech, graphic violent/sexual content, child abuse content, etc., is that it obfuscates the 'marketplace of ideas' within online platforms. The 'marketplace of ideas' is based on the principle that ideas should be freely able to compete with one another, with the best idea beating out its competitor for time, space, and resources in terms of popularity and reach (Waldman 2018). But, given the way social media algorithms operate with the system of filter bubbles, echo chambers, and hyper-customisation, they serve to disrupt this 'free market of ideas'.

It is prescient to remember the words of Marshal McLuhan, who described media as "extensions of man", with electronic media paving the way for a 'global village' with no single and siloed physical space being replaced by interactions and seamless connections on a global level. He also made a grim assessment of 'digital space' paving the way for neo-tribalism, fuelled by aggressive violence and human power as the proverbial 'clash of civilisations', i.e., "intermingling, clash and the explosion of cultural norms" would be occasioned (Satell 2021). The ground zero of social media platforms and internet space is a stark reflection of this assessment. In order to fully unpack and comprehend the Disinformation Economy, it is important to understand the twin-fold economic direction as envisaged by the 'cost' of disinformation on the one hand and the 'profits' of disinformation on the other hand. It is mediated by another factor, which can be termed as the 'price' of disinformation. The interplay between these three variables in the disinformation economy has to be studied, researched, and regulated if we are to curb the hazards posed by disinformation effectively (Fig. 4.2).

Fig. 4.2 The market for disinformation

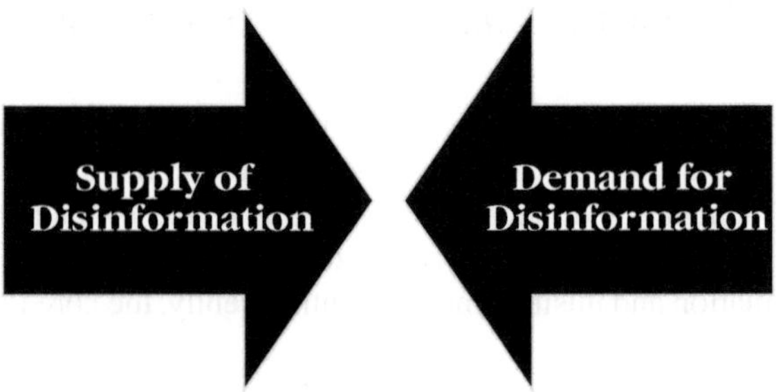

The economic forces of 'demand' and 'supply', which serve as the basis for understanding the dynamics of any market, apply to the market for 'disinformation' as well. Disinformation is a larger part of the 'information' market, which is concerned with the free availability rather than the accuracy of the 'information'. This is often the starting point of any discussion on 'information economics' (Taiji 2023). In the context of this discussion, we are concerned about two types of actors, viz., those who consume 'information' and those who produce 'information'. One significant aspect to be factored in the context of social media platforms is that the person often serves as both the supplier (content creator/sharer) and consumer (content consumer), which poses additional difficulties. Additionally, it can be seen that the decision to verify or fact-check any particular information before sharing the same entails weighing the cost of achieving certainty against the risk of sharing misinformation and disinformation (Cisternas and Vasquez 2022). Social media platforms and fake content producers benefit from continuous and uninterrupted engagement without pausing to recheck to spread the greater reach and profits, thereby disincentivising social media platforms from dis/disinformation regulation.

4.2.1 The Cost of Disinformation

The Disinformation Economy has two broad facets, viz., the cost of disinformation and the profits of disinformation. The cost of disinformation refers to the economic impact of disinformation and misinformation in terms of the monetary and economic loss incurred by countries, corporations, and people on this account. As per an estimate, the cost of disinformation is pegged to be around 78 billion dollars, which is the worldwide economic loss occasioned by the spread of 'fake news' in political, financial, and healthcare fields (CESIE 2022). Fake news and disinformation result in a stock market loss to the tune of 39 billion dollars each year (Brown 2019). Similarly, health misinformation losses and financial misinformation costs are estimated to be 9 billion USD and 17 billion USD annually. Another significant but often ignored component is the costs undertaken to counter disinformation and misinformation at the national and international level by governments, corporations, businesses, and social media conglomerates (Fig. 4.3).

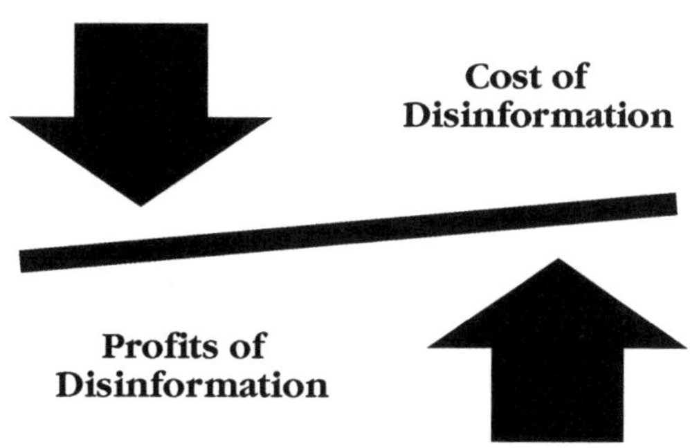

Fig. 4.3 Winner-loser model of disinformation Economy

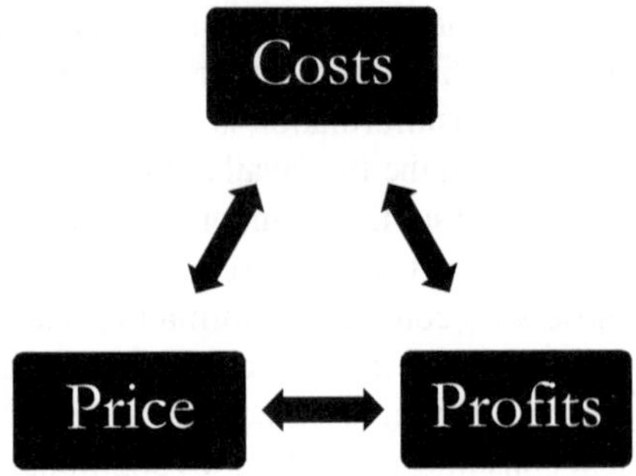

Fig. 4.4 The economy for disinformation

The 'disinformation' economy has created a Winner-Loser model which has paved the for two sets of actors, viz., the ones who profit out of peddling disinformation in so far as it has become a 'service' for sale and the ones who have to bear the cost (loss) of such disinformation. The peddlers of disinformation and misinformation are involved in profiteering while bleeding nations, corporations, brands, and other entities dry (Fig. 4.4).

Another interesting dimension for the dissemination and proliferation of disinformation and misinformation is what is termed the social costs of 'fake news' (Lawson 2023). Researchers have pinpointed that not sharing and engaging with 'fake news' has a social cost, thereby driving a hard bargain between spreading misinformation and having to contend with social exclusion (Lawson 2023). Tracing the evolutionary-biological urge for group membership as a necessary facet of our self-worth, scholars have concluded that peer pressure and conformity are important factors that play a major role in reinforcing the bond between people who share such 'fake news' content (Lawson et al. 2023). It has also been established that people who do not conform to the group's behaviour of sharing misinformation are subjected to social penalties (Lawson 2023).

As per a recent research study, there is a linkage between consumer preferences and 'fake news' and objectionable content, which indicates that consumers are less likely to engage and subsequently stop using a brand or product that is advertised next to fake, objectionable and inflammatory content (State of Digital Publishing 2019). Thus, being placed in advertisements next to such content can prove to be the proverbial 'death knell' for companies and brands. Recently, several reputed companies, including Disney, IBM, Apple, etc., decided to halt funding their ads through X (Twitter) on the allegation of their advertisements being found next to inflammatory anti-Jewish content (Mac et al. 2023).

4.2.2 The Price of Disinformation

In 2017, it was found that benignly named 'influence campaigns' could actually be orchestrated disinformation and misinformation campaigns that charge around $400,000 for a year-long fake news campaign, which includes the creation of fake profiles/accounts, writing false news stories and spreading them through/with fictitious followers (BBC 2017). The services offered include the creation of celebrities, discrediting journalists, and fuelling social unrest, including demonstrations and lobbying in connection with elections/political parties (BBC 2017). Disinformation campaigns lead to a variety of consequences ranging from maiming the prospects of electoral candidates, reputational damage to products or companies, manipulation of share prices with official social media presence serving to heighten the danger as well as ensuing harm before any damage control measure can be initiated. Disinformation campaigns typically rely on the triple-prong approach of utilising, manipulating, and even weaponising emotions, truth and repetition (Prevency, n.d.).

Another significant development is the growth and proliferation of technologies that render disinformation campaigns cheaper by the day. To put it in other words, the currency of 'truth' is becoming costlier than ever before, while the commodity of 'lies' is cheaper than ever before. For example, earlier disinformation campaigns were mounted through the use of 'useful idiots' and 'trolls-for-hire', entailing several thousand manual accounts, has been replaced by automated bot accounts and artificial intelligence-driven technologies. Recently, as a case in point, an AI-driven algorithm was deployed by a developer to produce Anti-Russian counter-propaganda in the form of tweets, articles, news content and journalists entirely fabricated by AI-backed tools to outline and underscore the dangers of AI-backed disinformation campaigns that can be mounted for as low as 400 USD (Knight 2023).

The primary reasons why disinformation campaigns are more easily made viable boil down to three cardinal reasons, viz., costs, anonymity, and credibility. With respect to costs (expenses) incurred, it is easier and cheaper to mount a disinformation campaign than a legitimate advertising campaign. As anonymity is ensured, it is easier to conceal a disinformation campaign rather than paid advertisements, and finally, the 'virality' phenomenon couches it with 'credibility' rather than advertisements (Gu et al. 2017). Disinformation campaigns targeting and aimed at discrediting a journalist are priced at $50,000 through systemic contradictory 'fake news', bombarding comments with negative comments (BBC 2017).

4.2.3 *The Profits of Disinformation*

The Disinformation Economy has created a series of stakeholders, including companies, corporations, individuals, and entities, that now offer 'Disinformation-as-a-Service (DAAS)'. Digital advertisements are also an important source of revenue for peddlers of disinformation and fake news. Disinformation services are offered under the garb of content marketing services through underground markets in China, Russia, Middle Eastern and European countries. Paid posts, reposts and promotions are also deployed in this endeavour (Gu et al. 2017). Similarly, 'likes', 'views' and 'shares' are also available for 'purchase' through such platforms. The cost of such pay-per-post 'promotions' masquerading as genuine 'reviews' or 'viewpoints' is abundantly found notwithstanding specific social media policies relating to 'sponsored posts' and 'paid promotions', including the requirements of clear indicators within the post itself. Disinformation is profitable in an economy where people's attention is 'bought' and 'solid' in millions and by the millions.

Intriguingly, the service of 'content takedown' is also offered as part of the disinformation economy and entails the takedown of 'negative content' relating to the payer for a hefty fee through many ways, including unscrupulous ones such as bribing the website administrator or hacking the website to remove such content (Gu et al. 2017, p. 18). Similarly, services such as online poll manipulation, sales of followers, crowdsourcing content, likes, and subscribers are also offered for commensurate prices. Toolkits that perform automated social media spamming are also offered for sale in Russia (Gu et al. 2017, p. 32). There are also sites that enable users to generate their own breaking news for free which can be used to subserve the propaganda and disinformation agenda with relative ease (Gu et al. 2017, p. 38).

Parallelly, there is tremendous scope for monetising and profit-making through the conduit of digital advertisements that are placed along with dubious content. Ad tech companies deploy algorithms that designate advertisements across the internet with little or no oversight by the actual brands themselves (Atkin 2023). But these advertisements, when placed alongside content of dubious nature, including the nature of disinformation and misinformation, create a self-serving double-benefit for such websites and platforms, creating a semblance of credence and additional respectability, which further enhances the growth of propaganda and disinformation campaigns. Thus, these platforms are able to derive profits both from the unwitting reputed brands and the unscrupulous benefactors of the propaganda and disinformation campaign while misleading the general public.

4.3 Part 2: Classifying the Colossal Conundrum

Disinformation, in its broadest sense, can concern, deal with, and afflict almost any aspect of human existence. Nevertheless, the phenomena and multi-headed hydra of disinformation have afflicted four segments/fields of human endeavour in comparison to all other aspects. This encompasses electoral politics, financial sectors, the study of science and, more importantly, the branch of medicine. It is also important to note that in the case of the latter two segments, it targeted the scientific and medical community directly, but its relation to the general public sowed doubts and confusion in their minds (Fig. 4.5).

The perils of artificial and synthetic manipulated content in misleading people, impairing economies and fracturing societies in manifold ways in the course of the upcoming 2 years have been specifically highlighted by the WEF Global Risks Report 2024 (WEF 2024, p. 18). Deepfakes, machine learning, artificial intelligence, the Internet of Things (IoT), predictive algorithms, etc., represent the new frontiers that pose novel challenges in the fight against disinformation.

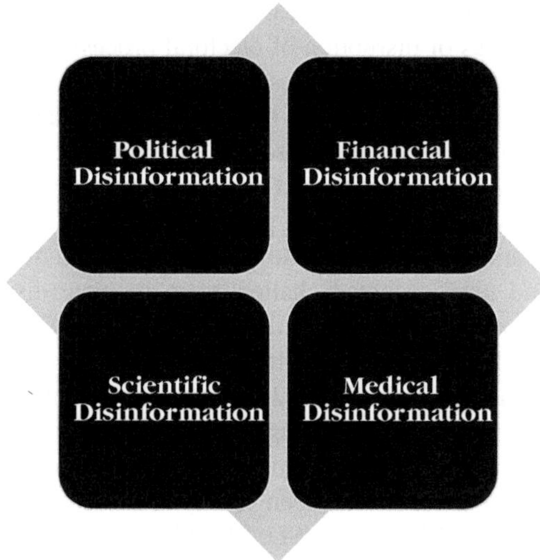

Fig. 4.5 Classification of disinformation

4.3.1 The Economics of Electoral Disinformation

Democracies around the globe are set for a historical milestone in the year 2024. It is estimated that around 70 countries of the world, encompassing around 3.7 billion people accounting for almost half the earth's population, including India, the USA, the UK, and Russia, will have their national elections this year (Rizzi 2024; Prasannan 2024). In this backdrop, the World Economic Forum (WEF) Global Risks Report 2024 has flagged the impact of disinformation and misinformation as the biggest threat to the economy in the short term, given that many global nations are set to go to polls (Elliott 2024). Disinformation and misinformation have the propensity to undermine the newly elected governments' legitimacy (WEF 2024, p. 8). The problem is accentuated by the rapid progress of AI technologies that have become intertwined with the creation of disinformation around the world. These 60 countries represent 60% of the global GDP and encompass India, the EU, the USA and the UK and the cascading effect on the global economy considering the cost-of-living crisis coupled with burgeoning disinformation and increasingly 'divided' societies (WEF 2024, p. 13).

The impending elections also cover countries such as Bangladesh, South Sudan, Pakistan, Tunisia, Taiwan, Venezuela, Ukraine, etc., which have faced several crises, including domestic unrest, riots, food shortages, human rights violations and war in the recent past. Hence, the trajectories and future of many of these countries will be determined by the outcome of this mammoth electoral exercise. As per an estimate, it is projected that a repetition of such a large number of countries going to polls in the same year is expected only in the year 2048 (Prasannan 2024). Given that many of these countries are liberal democracies that are expected to have elections in the ensuing few months, the menace of disinformation and misinformation poses a significant threat to them. The Global Risks Report 2024 underscores the chances of disruption of electoral processes through misinformation and disinformation with the propensity of civil unrest triggered by growing distrust of media and government. It also reiterates that there is an increased risk of civil liberties and human rights being repressed and eroded in the attempt to scuttle the spreading of such disinformation and misinformation (WEF 2024, p. 18).

Around three billion people are expected to cast their ballots in the upcoming 2 years; the problem of the rapid proliferation of generative AI in the mix creates new and unprecedented challenges to regulate disinformation, posing a dual risk of quantitative and qualitative nature (Chan 2024). Generative AI poses a quantitative challenge in so far as its capabilities to mass produce synthetic content without human efforts and a qualitative challenge with greater and more sophisticated deepfakes that are increasingly difficult is not impossible to flag by the layperson as being fabricated or manipulated (Rizzi 2024). It is to be noted that the deployment of 'fake news' has been in vogue for quite some time, as research showed that around 140 million USD, 34 million USD and 20 million USD were spent in India (2019), Brazil (2018) and Kenyan elections (Brown 2019).

The rampant spread of disinformation and misinformation leads to a growing distrust in government, which leads to further political polarisation. More importantly, not only are people's views, ideologies and outlook increasingly polarised, but people's perception of reality is undergoing polarisation, affecting social cohesion and mental health (WEF 2024, p. 20). The increasing spread of disinformation gives a free license to authoritarian norms in the digital sphere that range from censorship and surveillance to decreased internet freedom, paving the way for the government to have the sole prerogative of determining the 'truth', thereby minimising plural voices, viewpoints and dissenting ideologies.

There have been several instances of governments, particularly in authoritarian regimes, targeting human rights activists, opposition leaders, civil liberties groups, NGOs, journalists, etc., on the ground of spreading misinformation and disinformation. With respect to India, misinformation and disinformation have been flagged as the single biggest threat considering the nation's General elections scheduled in April–May 2024, coupled with the unprecedented internet and social media penetration reached as of date, making it extremely susceptible to social media manipulation (WEF 2024, p. 106). False social media narratives exacted a heavy toll in Nigeria, heightening tribal differences and leading to ethnic tensions, resulting in violence and disenfranchisement (Olaigbe 2023).

The Fake News Triangle revolves around the unholy trinity of social networks, motivations and allied tools/services. Political motivations and propaganda are two of the biggest driving forces behind the dissemination of disinformation and misinformation. Given that political disinformation and propaganda through social media are turning out to be one of the cheapest, most effective and precise ways to target a demographic group, there is no doubt a steady rise in the deployment of political disinformation for the years to come. Disinformation campaigns, including instigating a street protest, discrediting journalists, and manipulating a decisive course of action such as a referendum, elections, etc., can be carried out for prices ranging from 55 K USD to 400 K USD (Gu et al. 2017).

4.3.2 The Fiscal Costs of Financial Disinformation

In the recent past, corporate conglomerates have become the target of misinformation and disinformation campaigns. Research studies indicate that this results in additional costs being incurred to protect themselves from such malicious campaigns, including $9 billion spent on reputation management and $3 billion a year for platform safety efforts (Brown 2019). Similarly, the loss of brand dollars advertising next to fake news is pegged at $235 million.

The Global Stock Market is often at the receiving end of several disinformation and misinformation campaigns, with stock price falls occasioned by social media trends and news. For example, social media sharing of incidents such as fist fights and forcible removal of passengers from flights led to a fall in stock prices of United Airlines and Spirit Airlines (Gu et al. 2017, p. 54).

Many actors are involved in such disinformation campaigns, including disgruntled former employees, business competitors, lobby groups, trolls-for-hire, dubious PR companies, criminals or computer programs used by humans such as automated bots (Prevency, n.d.). Disinformation campaigns are often accompanied by boycott calls and trends calculated to bring significant economic loss and have the tendency to cause irreparable reputational harm and damage to the targeted brands/companies.

Another recent development pertains to the display of advertisements by reputed global brands, such as websites and internet platforms, that are regularly known to peddle misinformation and click-bait-type content. Thus, it makes several ad tech companies and big-name corporations unwitting partners in funding domains that peddle disinformation and misinformation. As per the Global Disinformation Index Study of 20,000 websites that publish information, it has been estimated that around $235 million is spent annually by ad tech companies on such websites (Tardáguila et al. 2019).

With global digital advertising valued at 600 billion USD, many companies channelise advertisements through a variety of channels across the open web, with several thousand websites hosting monetised advertisements, making it susceptible to be posted alongside shady content in the nature of disinformation and online fraud. A study by the World Federation of Advertisers estimates digital advertising to become the second-highest source of income for organised crime after the drug trade by the year 2025 (Atkin 2023).

4.3.3 The Sooq for Scientific Disinformation

Scientific disinformation and misinformation have plagued human civilisation from time immemorial. It ranged from beliefs such as flat earth, geo-centric model, existence of aliens/mystical creatures, etc. In the modern context, it takes a more sinister form with an admixture of half-baked knowledge, pseudoscience and vested interests leading to manifestations such as vaccine denial/hesitancy, climate change denial, false claims regarding environment degradation, fossil fuels, genetically modified organisms (GMO), tobacco, etc. Thus, it refers to the intentional dissemination of information relating to science that is untrue or false.

Parallelly, there is also the problem of misreporting, misstating, inaccurate representation, overstating and exaggeration of scientific research when it is communicated to the public, particularly by journalists and online platforms. The rigour and denseness of primary scientific literature, which may make it almost directly inaccessible to the general public, warrants the intervention of mediators such as journalists, other media personalities and communication intermediaries who may not fully grasp the nuances of the scientific study.

Science as a discipline presupposes verifiable empirical data subject to independent verification premised on rigorous peer review by experts and other qualified professionals in the field on the data, methods and research in question. Thus, there is an expectation of objectivity, rationality, accuracy, and verifiability, which are the cornerstones of scientific inquiry. Thus, any scientific claim should be backed by hard evidence and measurable data. Scientific misinformation and disinformation are an antithesis to all these accepted notions and understanding of the broad scientific community.

Scientific misinformation is defined as "publicly available information that is misleading or deceptive relative to the best available scientific evidence or expertise at the time and that counters statements by actors or institutions who adhere to scientific principles without adding accurate evidence for consideration" (Southwell et al. 2022). Misinformation and disinformation, particularly relating to science, sweep aside scientific propriety, facts, evidentiary support and sophistication to pander to flawed public perceptions and sentiments driven by emotions, falsehoods and fabrications. Tom Nichols speaks of the 'death of expertise' as a challenge that both questions and undermines the role and relevance of 'experts' with the term itself used in a contemptuous manner to justify ignoring their advice (Hopf et al. 2019). This cataclysmic shift is perhaps the single biggest challenge that has to be grappled with in the current age.

Social media is now the fulcrum around which conversations, debates and discussions relating to scientific advancements, discoveries and inventions gain traction and popularity among the general public. It is also interesting to note that the spread of scientific disinformation and misinformation is rampant in the so-called 'first-world' developed nations such as the USA, countries in the West including Europe, and the so-called 'developing and underdeveloped' countries. There is also substantial division across political spectrums in relation to false information relating to vaccines, climate change and other such scientific misinformation/disinformation.

Climate change denial, which refers to the practice of denying, understating or misrepresenting the causes, effects and ramifications of human-induced pollution and environmental degradation on the earth and its environmental impact, including climate change, is another rampant problem. Climate denial undermines the global efforts to combat anthropogenic climate change and is driven by corporate conglomerates such as fossil fuel giants that fund, lobby and promote such deliberate disinformation and misinformation masquerading as 'scientific' analysis with a view to protect narrow business interests and to undo attempts at legal regulation. It creates unwarranted doubts and confusion in the minds of the general public, thereby misleading and deceiving them of the potential harms occasioned by unabated human action on the environment (Holden et al. 2017). It is also sometimes referred to as 'climate change scepticism'.

Social media platforms and some traditional media houses are involved in this attempt, which often entails crony capitalism, corrupt corporations and unethical research practices. A similar lobby also exists in the context of tobacco production companies that carry out research relating to their health effects. The scientific disinformation problem has begun afflicting the scientific community as well, with a growing percentage of scientific articles being retraced on account fraud, indicting the flaws and weaknesses of the peer review system. The triple interaction between scientific advancement, reputational rewards and financial returns is considered to be the reason behind this growing trend of problematic scientific tendencies within the publishing ecosystem that encourages distortion and exaggeration of results in the bid for new grants, promotions and distinctions (Hopf et al. 2019).

4.3.4 The Marketplace for Medical Misinformation

Medical disinformation and misinformation have proved extremely challenging with social media's advent. Historically, there has been a trend of medical misinformation and disinformation concerning various diseases, cures, vaccines and medications. This has been aggravated in the context of the global COVID-19 pandemic that witnessed an unprecedented festering of false information, leading to what experts termed a COVID-19 Infodemic. Reports have pinpointed the exact financial and health-related impact of such false information circulated in the backdrop of the pandemic. For example, a Canadian study by the Council of Canadian Academies estimated that misinformation affected COVID-19 infections, resulting in the loss of 2800 human lives and $300 million in hospital expenditure in November 2021 in Canada (Major 2023). The study estimated that around 2.35 million Canadians skipped vaccination during the intervening period because of widespread misinformation, which would have prevented about 200,000 infections, 13,000 hospitalisations and 3500 ICU admissions (Weeks 2023).

Vaccine hesitancy and surrounding misinformation are part of the larger and pre-existing anti-vaccine movement and conspiracy that gained greater traction in the backdrop of the COVID-19 pandemic, particularly in the United States of America. These erroneous claims contest vaccines' safety, necessity and effects contrary to established medical science and research. Such erroneous and misleading claims in the backdrop of the pandemic are estimated to have resulted in 319,000 deaths between January 2021 and April 2022 and also resulted in additional costs of $50 million–$300 million loss due to hospitalisations, long-term illness and economic work loss.

Similarly, the role of registered dietary influencers in the promotion of diet soda, sugar and other supplements in tandem with their corporate manufacturers through social media platforms such as Instagram and TikTok has raised a lot of concerns about the ethical propriety and legal validity of these actions (Wang 2023). The lack of accurate information or worse, false information has the potential to cause physical, nutritional and health harm to individuals, as evidenced by the 'vaccine hesitancy' among parents in Europe and its disastrous effect on children triggered by the fabricated report relating to the MMR vaccine being the cause of autism resulted in unprecedented levels of measles infection in Europe in 2018 (Hopf et al. 2019).

4.4 Regulating the Market for Disinformation

The question of regulation of the market for disinformation necessitates technological and legislative intervention, which leads to allied concerns of censorship and restrictions on free speech. Nevertheless, economists have devised newer methods to reduce 'disinformation' and 'misinformation' in social media other than the traditional solutions such as censorship, fact-checking and education. This alternative model explores ways to enhance the quality of information disseminated without imposing specific responsibility on entities in regulating content and flagging content as true or false. It proposes laying down limits and caps on the reach of messages without being overly restrictive on the internet users. It is akin to WhatsApp, and Facebook limits the number of people and groups to which messages can be sent by capping it to five in total (Khanna 2022).

The predominant model of social media regulation revolves around the self-regulation theory. Social media platforms have their own policies, frameworks, and user policies based on which they are entitled to restrict, remove and prevent usage and content posted within their platforms for violations of terms and conditions. Social media platforms have also backed and supported both independent and in-house fact-checking mechanisms by specialist employees, as well as advanced algorithm-driven tools with the help of AI, natural language processing and network analysis (Daan 2021). The problem of disinformation cannot be viewed in isolation and forms part of a trinity of internet trends entailing citizen surveillance and cyber-warfare by government, non-state and other entities for various reasons (Daan 2021).

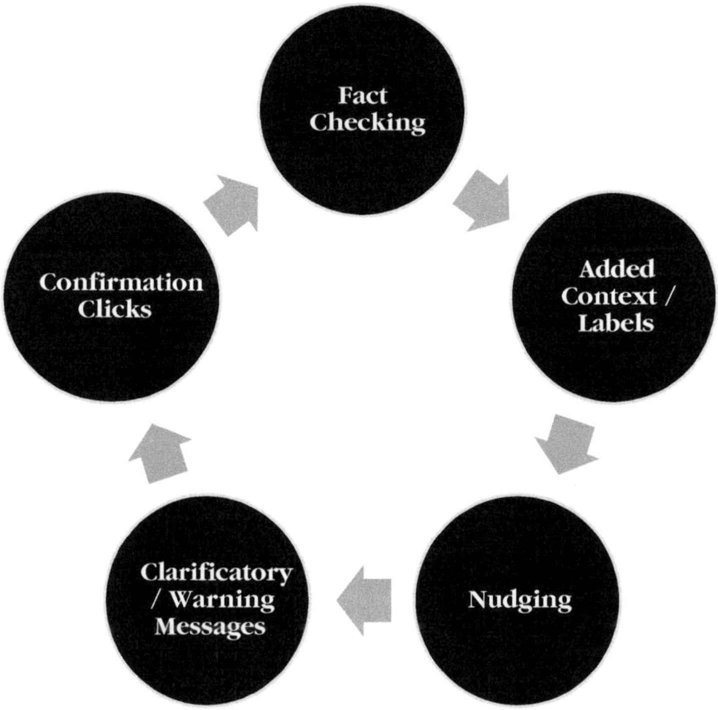

Fig. 4.6 Interventions for disinformation Regulation

Research and empirical studies show that one of the most significant ways to check misinformation is by warning users about the prevalence of false news before they share it. Anti-misinformation policies involve fact-checking, confirmation clicks, and clarificatory messages to prevent the spread of false information. These measures double-check the accuracy, relevance, impact, and cost of sharing (Guriev et al. 2023). Misinformation and disinformation can be countered by imposing monetary penalties in civil suits. Fox News paid 787.5 million USD as part of a settlement claim to Dominion Voting Systems for spreading false information about electoral fraud during the 2020 presidential elections due to their malice and reckless disregard for the truth. (Jones 2023) (Fig. 4.6).

4.5 Conclusion and Suggestion

In conclusion, it is often said that "bad economics cannot make good law". Addressing the concerns of disinformation and misinformation in social media requires a nuanced understanding of the economic factors that incentivise the market for disinformation and misinformation. Any regulation that disincentivises disinformation or, at the very least, is able to stop the incentivisation of disinformation would go a long way in curbing the market for 'disinformation'. One interesting suggestion and proposal relates to the imposition of a punitive and additional tax burden on corporations, entities and platforms, particularly social media platforms, that profit out of the disinformation economy. It is argued that the levy of such a tax on disinformation would hinder the rapid dissemination of 'fake news' and would compel platforms to take corrective measures. The advent of generative AI also creates a new multi-pronged hydra of disinformation and misinformation. Given the ethical, privacy and legal concerns surrounding the misuse of generative AI and its propensity to create Deepfake videos, audio and other synthetic content, there is an urgent need to regulate its usage. A suitable example can be the mandatory requirement of an AI watermark in all content created with the help of AI as mandated in China, which may help individuals flag AI-generated and AI-hallucinated content more easily (WEF 2024). This is one of the many initiatives that will help the market for disinformation revert to the original 'information economy'.

References

Atkin, C (2023, August 21) Are your ads funding disinformation? Harvard Business Review. https://hbr.org/2023/08/are-your-ads-funding-disinformation

BBC (2017, June 15) Prices for fake news campaigns revealed. BBC. https://www.bbc.com/news/technology-40287399. Accessed 24 Nov 2024

Brown E (2019, December 18) Online fake news is costing us $ 78 billion globally each year. ZDNET. https://www.zdnet.com/article/online-fake-news-costing-us-78-billion-globally-each-year/. Accessed 24 Nov 2024

CESIE (2022, January 31) What is the true cost of disinformation. Centro Europeo di Studi e Iniziative con sede a Palermo. https://cesie.org/en/youth/true-cost-of-disinformation-mega/. Accessed 24 Nov 2024

Chan K (2024, January 10) AI-powered misinformation is the world's biggest short-term threat, Davos report says. AP News. https://apnews.com/article/artificial-intelligence-davos-misinformation-disinformation-climate-change-106a1347ca9f987bf71da1f86a141968. Accessed 24 Nov 2024

Cisternas G, Vasquez J (2022, November 28) Breaking down the market for misinformation. Liberty Street Economics. https://libertystreeteconomics.newyorkfed.org/2022/11/breaking-down-the-market-for-misinformation/. Accessed 24 Nov 2024

Daan H (2021) An analysis of fake news and its effects on the economy and society. MPRA Research Paper No. 116027. https://mpra.ub.uni-muenchen.de/116027/. Accessed 24 Nov 2024

Elliott L (2024, January 10) AI-driven misinformation biggest short-term threat to global economy. The Guardian. https://www.theguardian.com/business/2024/jan/10/ai-driven-misinformation-biggest-short-term-threat-to-global-economy. Accessed 24 Nov 2024

Gu L et al (2017) The fake news machine how propagandists abuse the internet and manipulate the public. Trend Labs Research Paper. https://i2.res.24o.it/pdf2010/Editrice/ILSOLE24ORE/ILSOLE24ORE/Online/_Oggetti_Embedded/Documenti/2017/06/21/wp-fake-news-machine-how-propagandists-abuse-the-internet.pdf. Accessed 24 Nov 2024

Guriev S et al (2023, December 10) Evaluating anti-misinformation policies on social media. CEPR VOX EU. https://cepr.org/voxeu/columns/evaluating-anti-misinformation-policies-social-media. Accessed 24 Nov 2024

Holden, W, Nadeau, K, Porio, E (2017) Ecological liberation theology: Faith-based approaches to poverty and climate change in the Philippines. Springer.
Hopf, H, Krief, A, Mehta, G, Matlin, SA (2019) Fake science and the knowledge crisis: Ignorance can be fatal. Royal Society Open Science 6(5): 190161. https://doi.org/10.1098/rsos.190161
Jones D (2023, April 19) Fox News' pricey settlement shows cost of misinformation, local experts say. FOX 8. https://www.fox8live.com/2023/04/19/fox-news-pricey-settlement-shows-cost-misinformation-local-experts-say/. Accessed 24 Nov 2024
Khanna S (2022, August 23) Economists have a method for reducing fake news on social media. Duke Today. https://today.duke.edu/2022/08/economists-have-method-reducing-fake-news-social-media. Accessed 24 Nov 2024
Knight W (2023, August 29) It costs just $400 to build an AI disinformation machine. Wired. https://www.wired.com/story/400-dollars-to-build-an-ai-disinformation-machine/. Accessed 24 Nov 2024
Lawson A (2023, April 3) The social costs of not sharing fake news. Knowledge. https://knowledge.insead.edu/strategy/social-costs-not-sharing-fake-news. Accessed 24 Nov 2024
Lawson MA, Anand S, Kakkar H (2023) Tribalism and tribulations: the social costs of not sharing fake news. J Exp Psychol Gen 152(3):611–631. https://doi.org/10.1037/xge0001374
Mac, R, Barnes, B, Hsu, T (2023, November 17). Advertisers flee X as outcry over Musk's endorsement of antisemitic post grows. The New York Times. https://www.nytimes.com/2023/11/17/technology/elon-musk-twitter-x-advertisers.html
Major D (2023, January 26) COVID-19 misinformation cost at least 2800 lives and $300M, new report says. CBC News. https://www.cbc.ca/news/politics/cost-of-covid-19-misinformation-study-1.6726356. Accessed 24 Nov 2024
Olaigbe O (2023, March 22) Disinformation has a human cost in Africa's Largest Democracy. Inkstick Media. https://inkstickmedia.com/disinformation-has-a-human-cost-in-africas-largest-democracy/. Accessed 24 Nov 2024
Prasannan R (2024, January 14) A universal adult festival. The Week. https://www.theweek.in/columns/prasannan/2024/01/06/a-universal-adult-festival.html. Accessed 24 Nov 2024
Prevency (n.d.) What is a disinformation campaign. Prevency Insights. https://prevency.com/en/what-is-a-disinformation-campaign/. Accessed 24 Nov 2024
Rizzi A (2024, January 16) With billions to vote in 2024, world economic forum ranks disinformation as top short-term threat. El País. https://english.elpais.com/international/2024-01-16/with-billions-to-vote-in-2024-world-economic-forum-ranks-disinformation-as-top-short-term-threat.html. Accessed 24 Nov 2024
Rosenfield S (2019) Democracy and truth – a short history. University of Pennsylvania Press
Satell G (2021, September 5) The economics of disinformation. Digital Toronto. https://digitaltonto.com/2021/the-economics-of-disinformation/. Accessed 24 Nov 2024
Shumaker D (2019, October 15) Supply and demand: the economy of disinformation. News Breaks. https://newsbreaks.infotoday.com/NewsBreaks/Supply-and-Demand-The-Economy-of-Disinformation-134666.asp. Accessed 24 Nov 2024
Southwell BG et al (2022) Defining and measuring scientific misinformation. Ann Am Acad Pol Soc Sci 700(1):98–111. https://doi.org/10.1177/00027162221084709
State of Digital Publishing (2019, June 18) Study: Consumers reject brands that advertise on 'fake news' and objectionable content online. https://www.stateofdigitalpublishing.com/monetization/consumers-reject-brands-that-advertise-on-fake-news/
Taiji H (2023, February 1) An economic theory of disinformation. MPRA Paper No. 116177. https://mpra.ub.uni-muenchen.de/116177/. Accessed 24 Nov 2024
Tardáguila C et al (2019, September 26) How ad tech companies fund misinformation. Poynter. https://www.poynter.org/fact-checking/2019/how-ad-tech-companies-fund-misinformation/. Accessed 24 Nov 2024

Waldman AE (2018) The marketplace of fake news. Univ Pa J Constit Law 20:845–870. https://scholarship.law.upenn.edu/jcl/vol20/iss4/3. Accessed 24 Nov 2024

Wang M (2023, December 17) Health misinformation online is extracting a dangerous cost. Here's how to identify it. Scroll. https://scroll.in/article/1060627/health-misinformation-online-is-extracting-a-dangerous-cost-heres-how-to-identify-it. Accessed 24 Nov 2024

Wardle, C, Derakhshan, H (2018) Thinking about 'information disorder': Formats of misinformation, disinformation, and mal-information. Journalism, 'Fake News' & Disinformation 43–54.

World Economic Forum (2024, January 10) Global risks report 2024. https://www.weforum.org/publications/global-risks-report-2024/

Weeks C (2023, January 26) Vaccine hesitancy, COVID conspiracies led to thousands of deaths, high cost for health care system, report finds. The Globe and Mail. https://www.theglobeandmail.com/canada/article-report-the-high-human-cost-of-misinformation/. Accessed 24 Nov 2024

B. Karthik Shiva is Assistant Professor of Law at the Tamil Nadu Dr. Ambedkar Law University (TNDALU), Chennai, and Doctoral Scholar at the National Law Institute University (NLIU), Bhopal, where he is pursuing a Ph.D. on the impact of fake news on democracy and human rights in India.

Karthik completed his B.A. LL.B. (Hons) at the School of Excellence in Law, TNDALU, in 2017, and an LL.M. in Constitutional Law and Human Rights in 2019, where he was awarded the University Gold Medal. His LL.M. thesis focused on data privacy in the digital era. He has also qualified for the National Eligibility Test (NET) and Junior Research Fellowship (JRF).

He has taught at multiple institutions, including Andaman Law College, the VIT School of Law, and TNDALU, where he serves as Assistant Professor. His academic interests include constitutional law, human rights, digital privacy, fake news, and disinformation.

Karthik has made significant contributions to legal scholarship, publishing articles and chapters on topics including privacy, fake news, and cyber law. He co-edited the book *Law and Agriculture: National and International Perspectives* (2022), and his work has been featured in prestigious journals. He has received numerous accolades, including the University Gold Medal for his undergraduate and postgraduate studies, the Best Research Paper Award at an international IPR conference, and the Best Intern Award during his internship at the NHRC (2015–2016). Karthik regularly presents his research at national and international conferences on the regulation of fake news, climate change, and legal education. He serves as a peer reviewer for the *ILI Law Review Journal* and as Faculty Co-editor of *Lex Navis Magazine* at the Andaman Law College. He is also a member of the International Association of Constitutional Law (IACL–AIDC). His research continues to explore the evolving relationship between law, technology, and human rights.

Chapter 5
Surrogate Advertising in India

Aman Amrit Cheema and Ankur Taya

Abstract Surrogate advertising, a practice born out of necessity, has its origins in India's strict regulations governing the advertising of specific goods. Industries such as tobacco, alcohol, and betting applications face severe restrictions on direct advertising due to their legal, societal, and health implications. In response, businesses devised surrogate advertising as a strategic means to indirectly promote these products while ostensibly endorsing different, seemingly unrelated products. The first part would be a brief introduction to the concept of surrogate advertising. The paper would then trace the historical evolution of surrogate advertising in India. Similar to the tobacco industry, many alcoholic beverage brands embraced surrogate advertising. And, the non-alcoholic products like water, soda, or merchandise became vehicles for indirect promotion, allowing companies to maintain brand visibility despite advertising restrictions. With time, industries including gaming, lottery services, aerospace, and defence also started using surrogate approaches to get around restrictions on advertising while preserving their brands. In the third part, the paper will identify and analyse the prevalent forms of surrogate advertising across various industries. There are various forms of surrogate advertising that include product substitution, lifestyle and experience marketing, nationalism and patriotism, and so

The original version of the chapter has been revised. A correction to this chapter can be found at https://doi.org/10.1007/978-981-95-3361-9_15

The chapters in this book are presented in the "International Conference on Legal and Economic Synergy for an Equitable Society, organised by the University Institute of Laws at Panjab University Regional Centre, Ludhiana, India on February 1st, 2024".

A. A. Cheema
University Institute of Law, Panjab University Regional Centre, Ludhiana, Punjab, India
e-mail: aamanrit@pu.ac.in

A. Taya (✉)
Department of Laws, Panjab University, Chandigarh, India
e-mail: adv.ankurtaya@gmail.com

A. Virk et al. (eds.), *Law, Economy, and Society: Insights into Legal and Economic Issues in India*, Accounting, Finance, Sustainability, Governance & Fraud: Theory and Application, https://doi.org/10.1007/978-981-95-3361-9_5

on. Furthermore, the regulatory landscape governing surrogate advertising in India will be examined in detail. In the last part, the researchers would provide insights into challenges and prospects associated with regulating surrogate advertising.

Keywords Advertisement · Surrogate · Regulations · Restriction · Economy

5.1 Introduction

> Advertising is legalised lying. (H.G. Wells)[1]

Promoting a product or service with extensive exposure to the public is a useful method of conveying its value. It is appealing enough to enable quick connections between the product and customers located anywhere in the world (Das 2016). Nonetheless, some products are prohibited from being advertised in any way by various legislations in India. These goods are typically seen as detrimental to society and have highly addictive qualities. That is why their advertisements are banned. Interestingly, there are some companies that resort to what H.G. Wells said was legalized lying. Hence, the companies found a new way to advertise the banned products. This new phenomenon is 'Surrogate Advertising'.

The term 'surrogate' is defined as a 'substitute', i.e., anything which takes the place of another. Surrogate advertising is the practice of using one product to advertise another. In particular, the goods for which it is illegal under law to advertise directly in mass media (Das 2016). As an instance, the law forbids the promotion of cigarettes and alcohol.

The proprietors of cigarette and alcohol businesses replicate the packaging of their own products in an attempt to use the brand equity to sell the prohibited items. When consumers look at these advertisements, they get the inference that the commercial is of the banned product and not the product being advertised. As a result, the businesses are able to preserve their brand recognition even when the product that was initially being advertised is one for which media outlets are unable to air advertisements.

This has been affirmed by Dr. S.G. Khawas Patil and others in their research titled "Surrogate Advertising: A Successful Marketing Strategy for Liquor, Whisky Products" (Khawas Patil 2011). This research paper examines the customers' awareness of surrogate advertisements and whether or not the things featured in these commercials are actually being sold in the market or not. They concluded that the maximum respondents can tell the intention behind surrogate advertisements. Take the instance of Music CDs, Royal Stag or Seagram Whiskey (both liquor brands), it is obvious that substitute goods do not pique the interest of either the manufacturer or the consumer. The reason is that these products are not easily available, and even when they are available, they are very less in number as compared to the main product. To put it briefly, the businesses are marketing the appropriate products and

[1] H.G. Wells' (2009) statement that "*Advertising is legalised lying*" serves as a **critical lens** to introduce the issue of surrogate advertising.

following the law. Therefore, using surrogate advertisements to market whisky and spirits items has been a successful tactic for them to bypass laws (Supra Note 6).

A growing ethical concern in Indian advertising is surrogate advertising. Media outlets are replete with surrogate advertisements. Be it social media, mass media, online gaming applications, websites, print media, and so on. At present, this tactic is used by many companies throughout the country. However, this phenomenon could be traced back to the early twentieth century.

5.2 The Beginning of Surrogate Advertisements

During the early twentieth century in Britain, drinking became a custom prevalent among men. Due to this habit of drinking, domestic violence increased immensely in those times. Therefore, the British women started protesting against the liquor brands that advertised their products publicly. Ultimately, this put liquor companies on the defensive as they had to alter their methods of advertising. The companies that were facing backlash began marketing soda and juices while secretly selling alcohol by employing surrogate marketing techniques.

The Cigarette Advertising Act, 1975, which was passed by the Indian government in 1975, marked the debut of surrogate advertisements for tobacco companies in the country. The tobacco companies were required by this Act to include health warnings on the packaging and in advertisements for cigarettes. Following the Act's enactment, the Indian government and the health ministry persisted in advocating for more stringent regulations on the promotion of alcohol and tobacco products. The government outlawed tobacco and alcohol advertisements altogether in 1995. The Ministry of Health conducted a significant investigation before enforcing the prohibition. They discovered that drinking alcohol and smoking cigarettes both negatively impact one's health. Furthermore, according to the Indian government, these products specifically damage people's mental health, making them lethargic and uninspired. These elements working together eventually resulted in a prohibition on the products' advertising across all of its media channels (Singh 2015). However, sales of tobacco and liquor increased exponentially as the population grew. Consequently, businesses were compelled to look for alternate forms of advertising, which ultimately resulted in the development of surrogate advertising in India (Singh 2015).

The Indian tobacco firm 'Azad Bidi' sponsored an international cricket match in India, marking one of the first instances of surrogate advertising. Similar results from large-scale surveys conducted in India demonstrated that alcohol advertisements directly impacted the purchasing decisions of Indian consumers. After the Cable TV Regulation Act, 1995, was passed, it was forbidden to advertise cigarettes and alcohol. This was the period when many companies began to use counterfeit advertisements to promote their products, to show that they were still available in the market (Mehta and Nanda 2003). It's also a sort of proxy advertising when a spirits brand is used to fund sporting or cultural activities. In 1999, the Wills Cigarette, which was manufactured by ITC, sponsored the Indian cricket team, and the logo was printed on the sports apparel of the team. One Non-Government Organisation (NGO) named Voluntary Health Association of India filed a petition in the Delhi High Court seeking a ban on the same. But in 2001, Wills voluntarily ended its Indian cricket team's sponsorship while the court proceedings were still pending, stating that it did not want to obstruct the government's sincere attempts to outlaw tobacco products and associated items (Mehta and Nanda 2003).

In 2004, the Government of India released a 'Report on Tobacco Control in India' (Swasticharan et al. 2022) that showed that the expenditure on advertising by the tobacco companies has increased after the government outlawed their advertisements (Prohibition of Advertisement and Regulation of Trade and Commerce, Production, Supply and Distribution Act 2003). Before the ban by this 2003 Act, the cigarette advertising budget was Rs. 3000 million per year. However, it rose by 500 million rupees following the Act's enactment. After the ban, there was a trend among the cigarette and tobacco companies to sponsor the sports pages of various newspapers. Additionally, they began to portray tobacco chewing in their commercials, linking it to Indian rituals. While local bidi businesses resorted to advertising on transport trucks, international cigarette firms used to compete with one another in billboard advertising.

In the first decade of the twenty-first century, the era of the tobacco and cigarette industries' battle for the rights to sponsor different sporting and cultural events had started. A musical concert featuring the biggest names in vocal and instrumental music was sponsored by the cigarette brand 'Charms'. In a similar vein, the makers of gutka, 'Manikchands', sponsored the widely watched. Filmfare Awards event in India (*Supra Note 9).* And, these tussles of advertising led to a surge in surrogate advertisements in India.

5.3 Surrogate Advertising in India

Advertising works in a market setting where a number of factors, including customer requirements, corporate interests, and governmental laws, are at play. When it comes to informational content and persuasiveness, it is a formidable force. But increased prominence also brings with it more criticism and controversy. This is true because advertising, as a persuasive communication tool, affects people's choices, values, and lifestyles. For this reason, businesses utilize fictitious advertisements to draw attention to their products, whose broadcast advertisements are prohibited (Dodrajka 2011). They also use brand extensions, which is a strategy of leveraging an existing brand name or identity to introduce a new product or product category. Instead of creating an entirely new brand from scratch, companies use the existing recognition of their existing brand to use it as a surrogate advertisement for their cigarette, tobacco, or other product whose advertisement is banned. Utilising the existing product's brand equity is the company's approach.

It is now imperative for businesses to use surrogate advertising to promote their goods. The tobacco industry in India, for example, is heavily investing in non-tobacco items and aggressively marketing them in all forms of media under the same brand name as its tobacco products. Certain Gutka brands, like Pan Parag, Goa 1000, and Simla, use surrogate advertisements for paan masala under the exact same brand name to get around the prohibition on tobacco advertising on television networks. In North India, 502 Pataka Beedi is a well-liked brand, and in TV commercials, 502 Pataka chai (tea) is advertised. On media platforms, a deluge of paan masala commercials has appeared. In its advertising, the business emphasises that paan masala is a non-tobacco product by utilising the identical brand name for other non-tobacco goods. High-end retail establishments in all major metropolises were marketing Passion greeting cards under the Gold Flake (famous cigarette brand) logo (*Supra Note 9).*

Another big player in this game is the alcohol industry. Bagpiper soda, Haywards soda, Derby special soda, Gilbey green aqua are all alcoholic brands advertising non-alcoholic products.

Royal Challenge, a premium liquor brand, advertises its golf accessories and mineral water. Kingfisher mineral water and Kingfisher Airlines are other examples of surrogate advertisement. White Mischief holidays, Imperial Blue cassettes and CDs, Teacher's achievement awards, Smirnoff cassettes and CDs, and so on are a few examples of surrogate advertisements that are displayed in print, electronic, and outdoor media. The brand name and symbol on these products are precisely the same ones that viewers had previously seen in commercials for alcoholic beverages. It was very disappointing that even the big alcoholic beverage corporations, like Seagram's and McDowell's, were venturing into new industries like sports equipment, mineral water, cassettes, and CDs, etc., to maintain their brand presence. Subsequently, it was discovered that the principal aim of these fictitious commercials was to endorse their spirits brands, which included vodka, wine, and beer. The brand extension is a way to get around the ban on advertisements.

The advertising and marketing of betting, gambling, and online gaming apps is another sector of the economy that is currently booming. After the COVID-19 pandemic, the number of betting and gambling applications has enormously increased. There are platforms like Betway, Dafabet, Parimatch, Stake, FairPlay, and so forth (Tanzila 2023). Talking about surrogate advertising, Samit Sinha, managing partner of a brand consulting agency, was apt in decoding the surrogate advertising. In his words

> If you try to ban something that is a popular human pursuit, especially an addictive one, people will find a way around. Simply banning something will not stop people from consuming it. When it comes to surrogate advertising, alcohol and tobacco brands were primarily the ones to do it. Kingfisher has an airline as well as an alcohol brand. So, the entire area of surrogate advertisements has completely blurred lines, and it is difficult to say whether it is ethical or not.

Cricket already had a religious-like following in India, and the Indian Premier League (IPL) has further added to the popularity of the game. However, there is no denying the tremendous growth in betting marketing. And, this is despite the fact that betting is illegal in most parts of the country and the advertisement of gambling is regulated in a complex manner.

Thanks to the availability of cheaper smartphones and even cheaper availability of the internet, the market is ever increasing. India has a very large sports fanbase and that is why the gaming applications that allow paid games of skill have become popular. Despite the complex legal structure, betting operators are now attempting to attract this fanbase and have been successful to a large extent. They have started using the term 'gaming' instead of betting or gambling. There are many applications that are available on smartphones that allow you to bet with one click. These companies are dodging the law and carrying on with their promotions on Live TV and hoardings, print and social media. Even the top players like M.S. Dhoni, Virat Kohli, Rohit Sharma and so forth all are promoting these gaming applications through their social media and by working in their advertisements (Mehrotra 2023).

On March 21st, 2024, the Government of India gave an advisory to endorsers and influencers on social media, social media intermediaries, and online advertisement intermediaries. The notification advised the social media influencers to refrain from publishing or broadcasting advertisements of online betting platforms and any product/service depicting these platforms in a surrogate manner. Furthermore, the online advertisement adversaries have been advised not to target such advertisements towards the Indian audience. (Advisory, Ministry of Information & Broadcasting, Government of India, 2021).

It cautioned the influencers and intermediaries to comply with the order, or penal action under appropriate statutes would be taken. Also, the Ministry has the power to remove/disable the social media posts/accounts that promote online betting and gambling.

There are legal loopholes with regard to surrogate advertising in India. The legal structure dealing with these advertisements is somewhat less stringent than it should be.

5.4 Law Relating to Surrogate Advertisement

The statutes relating to the surrogate advertising in India primarily pertain to the regulation of products like tobacco and alcohol. The relevant laws can be discussed under the following heads.

The Cable Television Networks (Regulation) Act 1995 (CTNA) and Cable Television Networks Rules, 1994 (CTNR)

In 2008, February 25th, in a notification revising the CTNR, the government declared that no advertisement that directly or indirectly encourages the manufacture, sale, or use of tobacco products, wine, alcohol, spirits, or other intoxicants would be allowed. However, after a year, the I&B ministry released a notification revising this rule to permit, subject to certain restrictions, the advertising of goods sharing a brand name or emblem with any tobacco or alcohol product, but with caveats. The caveats were that the visual of the advertisement not to depict the product in the advertisement to be from the list of prohibited products. There shouldn't be any references to the prohibited goods in the commercial. Additionally, TV networks that cover news and current affairs should cease airing any product advertisements that contain the brand name or emblem of a company that sells cigarettes, tobacco products, wine, alcohol, or other intoxicants, and they should carefully adhere to the updated guideline (Rahim 2022).

Cigarettes and Other Tobacco Products (Prohibition of Advertisement and Regulation of Trade and Commerce, Production, Supply and Distribution) Act, 2003 ("COTPA")

According to Section 5 of the COTPA, no one involved in the manufacture, distribution, or supply of cigarettes or other tobacco products may run advertisements that advocate for the use or consumption of cigarettes or other tobacco products in any way. However, the new trend appears to be that of surrogate advertisements. The brand name is removed from the tobacco packets to avoid being subject to the law, and it is only used in unrelated or non-tobacco/alcoholic products. But the point here is that the brand name so used will be etched in the memory of the common man as a brand of tobacco earlier available in the market and produced by a certain company. The COTPA regulations make it very clear that using a tobacco product's name or brand to market, promote, or advertise another product is forbidden under Section 5 since it is considered an indirect form of advertising.

The Advertising Standards Council of India ("ASCI")

ASCI is a self-regulatory voluntary organisation of the advertising industry in India. It ensures the protection of the interests of consumers. It checks that all the advertisements conform to its code of self-regulation, which requires advertisements to be legal, decent, honest, and truthful. Therefore, this council's vigilance on advertisements that do not follow the standards is up to the mark. Manisha Kapoor, CEO and Secretary General of ASCI, in relation to these types of advertisements, said:

> Surrogate advertisements are in breach of the law, as are direct advertisements in restricted categories. ASCI has been vigilant in processing surrogate advertising, as well as reporting direct advertising to appropriate regulators. Together, we aim to effectively eliminate deceptive advertising practices. (Afacs 2024)

As per its guidelines, the narrative or imagery of an advertisement should exclusively include the product that is being promoted, and there should be no mention, either directly or indirectly, of products that are prohibited. Surrogate advertisements remain a concern in restricted categories. ASCI processed 49 such advertisements where its guidelines were being violated. Among these advertisements, 36 were alcohol related, 12 were from betting platforms, and one was for pan masala.

There were discussions between the Department of Consumer Affairs (DCA) and ASCI in Mumbai in 2024. The consultation also involved the All India Gaming Federation (AIGF) and E-Gaming Federation. They all pointed out the immediate need to stop surrogate advertisements. The discussion also focused on the rampant use of celebrities in the ads of prohibited products, which needs to be controlled.

The Surrogate Advertisements (Prohibition) Bill, 2016

This bill has made the punishment regarding the surrogate advertisements somewhat stringent. According to Section 5, anybody who violates the terms of this Act will be penalized with a term of imprisonment that must not be less than three years but may reach ten years, as well as a fine that must not be less than three lakhs but may extend up to ten lakhs. Section 6 states that any person who was in charge of or accountable to the company, firm, or association at the time the offence was committed for the management of the association's or company's business will be considered guilty and will face legal prosecution and punishment. Significantly, Section 7 specifies that the provisions of this Act shall not interfere with, and instead shall be added to, any existing laws that now govern any of the matters covered by it.

5.5 The Flip Side of the Coin

In support of surrogate marketing, the industrial section has a stance. Everything, according to the alcohol companies, is in compliance with laws passed by the government. Alok Gupta of the UB group is of the opinion.

> If a brand has equity, why shouldn't it be allowed to advertise? Also, brand extension is an industry practice adopted by different categories. (Supra Note 12) (Mehta and Nanda 2003)

This means that they have stopped showing liquor products, and they are following the law. According to the Confederation of Indian Alcoholic Beverages Companies (CIABC), alcohol advertisements have to be allowed. The liquor, cigarette, and now betting applications companies are using the popularity of Bollywood, cricketers, and many other celebrities and sportspersons. The largest market for surrogate advertisements is, arguably, Bollywood. Almost all films had smoking scenes in them prior to the government outlawing such scenes (Gangopadhyay and Dutta 2015).

Under the following headings, a few instances of surrogate advertisements have been covered:

5.5.1 Case Study I: Royal Stag

Pernod Ricard India is the owner of the premium whisky brand Seagram's Royal Stag. This company, like other whisky brands, caters to the nation's young, self-assured, and progressive men. This whisky company has a catchphrase:

> It's your life, make it large.

The business has been employing creative strategies to emphasise its intended message ever since the government tightened the restrictions. The company has supported sporting events and other entertainment-related activities both domestically and internationally. To target a younger audience, they have engaged well-known celebrities, such as movie stars and well-known cricket players. A quick glance at the company's website reveals that Royal Stag has sponsored a number of different events. There is a photo of a studio that has the atmosphere of a pub; however, it is devoid of any alcohol bottles. It makes it quite evident that the concepts are disguised. The website Royal Stag Large Short Films serves as a platform for showcasing short films from both aspiring and seasoned industry directors who hope to "make it large." A platform for music promotion, Royal Stag Mega Music is linked to TV shows like Royal Stag MTV Unplugged and Royal Stag Mirchi Awards. Numerous clubs in the well-known Indian Premier League are sponsored by Royal Stag. Under the surface of their efforts to support talent across a range of disciplines, their spirits business is supported. Youngsters, the target audience, are vulnerable. (*Supra Note 24)*

5.5.2 *Case Study II: United Spirits*

Another company that has used surrogate advertising to further its primary goal is 'Royal Challenge'. This spirits brand is closely associated with the IPL franchise Royal Challengers Bangalore, and the 'United Spirits' is the owner of the whisky. The IPL post-match festivities are sponsored by 'Signature', another United Spirits brand. On one side, cricket is followed like religion in India and on other side the alcoholic beverages companies are promoting their product by associating them with the popular sport. They do aggressive marketing through which they tend to attract the youngsters. The IPL teams are sponsored so that they can make a connection with the younger lot. One of their products, "McDowell's No. 1 Yaari," is promoted, for example, and it depicts the journey of three young friends through challenges, joy, and mischief. The commercial, which celebrates friendship, finishes with the phrase "Asli yaari ka no.1 Ehsaas" with McDowell's No. 1 emblem displayed prominently on screen. Additionally, McDowell sponsors Mohun Bagan, a well-known football team in India.

Companies like these focus on youth, games and aspirations associated with them, to reach their objective of increasing the sale of their products whose advertisements cannot be done directly as the law of the land.

5.5.3 *Case Study III: Betting Advertisements*

The Public Gambling Act of 1867 regulates betting and gambling in India. Although it is a state subject, many states have followed this Act with their own amendments. Gambling is altogether banned and is allowed in some Casinos in Goa and Sikkim. And, betting is allowed only if it is a 'game of skill' like horse racing. If one goes through social media, TV, and print media, there is a plethora of advertisements for betting applications, even though betting and gambling is prohibited in many parts of the country.

In response to increasing instances of advertisements promoting illegal activities such as betting and gambling, the Central Consumer Protection Authority (CCPA) has issued a comprehensive advisory. The advisory emphasises the prohibition of advertising, promotion, and endorsement of unlawful activities prohibited under various laws. The companies do this under the guise of 'gaming'. They do not call these applications 'betting applications'; rather, they give them the name of 'online gaming'. The endorsements of such activities present considerable financial and socio-economic implications, especially to the youth. (Ministry of Consumer Affairs 2024) The advisory underscores the Ministry of Information and Broadcasting's efforts in issuing various advisories to media platforms, cautioning them against publicizing betting and gambling platforms. Online advertisement intermediaries have also been warned against targeting such advertisements toward the Indian audience. Talking about the celebrities promoting the betting activities, CCPA said,

> It has come to our attention that betting platforms are employing celebrities and influencers to endorse and promote their betting activities. Consequently, endorsement by celebrities gives an impression that indulging in such activity is acceptable. Engaging in the promotion or advertisement of online gambling and betting, given its unlawful status in the majority of the states, renders one equally liable for participating in an illegal activity. Hence, celebrities and influencers are advised to refrain from endorsing and promoting illegal betting and gambling activities.

Portraying 'betting applications' as 'gaming applications' is nothing but surrogate advertising.

5.6 Conclusion

In many ways, surrogate advertisements are dishonest and untrue in addition to being misleading. Companies employ this strategy because the laws currently in force only do a haphazard job of containing this threat. If the government and legislature genuinely wish to put an end to this threat, they ought to carry out their plans with greater vigour and without the usual breaks. Either the goods should be outlawed and taken off the market if the government is sincere about solving this issue or it can make some provision which makes it practically impossible for the companies to use circuitous methods to make use of their brand names and make a mockery of the legislation. Given the prevalence of surrogate advertising, now is the time to address the issue head-on.

More authority should be granted to the ASCI so that it can prosecute false and deceptive advertising. It keeps a careful eye on surrogate advertising, but in order for it to function effectively, provision for strict punishment should be provided, rather than only warnings and forcing the guilty channels to apologise. The Ministry of Consumer Affairs ought to develop educational initiatives to help the public comprehend surrogate ads and how corporations use them to trick them. Additionally, the advertising agencies must be well aware of the goods sold under the same brand as the advertisements they are pushing, and they ought to sue any agencies that create surrogate advertisements.

References

Afacs (2024) ASCI consults with the Department of Consumer Affairs on surrogate ads. https://www.afaqs.com/news/advertising/asci-consults-with-department-of-consumer-affairs-on-surrogate-ads. Accessed 15 Mar 2025

Das T (2016) A study on surrogate advertising. MBA Dissertation, Delhi Technical University

Dodrajka S (2011) Surrogate advertising in India. Management and Labour Studies 36(3):281–290

Gangopadhyay S, Dutta S (2015, June) Surrogate advertising on Satellite television: Indian experience. J Bus Manage Comm Res IV(XII). https://www.researchgate.net/profile/Soumya-Dutta-5/publication/301892248_SURROGATE_ADVERTISING_ON_SATELLITE_TELEVISION_INDIAN_EXPERIENCE/links/572b9b4508ae2efbfdbdda56/SURROGATE-ADVERTISING-ON-SATELLITE-TELEVISION-INDIAN-EXPERIENCE.pdf. Accessed 15 Mar 2025

Khawas Patil, Dr. S.G., Hurne LS (2011, June) Surrogate advertising: "a successful marketing strategy for liquor, whisky products". Indian Streams Res J I(V)

Mehrotra, A (2023) 'Advertising of betting websites in India' (Asian Racing Federation, 6 November 2023) https://www.asianracing.org/email/202311qb-advertising-of-betting-websites-in-india

Mehta PS, Nanda N (2003, May 23) Surrogate advertising- needed, a spirited attack. The Hindu Business Line. https://pradeepsmehta.com/surrogate-advertising-needed-a-spirited-attack/. Accessed 15 Mar 2024

Ministry of Consumer Affairs (2024) Food & Public Distribution. Central Consumer Protection Authority issues advisory on Prohibition of Advertising, Promotion, and Endorsement of IllegalActivities. https://www.pib.gov.in/PressReleasePage.aspx?PRID=2011879#:~:text=Betting%20and%20gambling%20are%20stri%20ctly%20prohibited%20under%20the%20Public%20Gambling,under%20the%20guise%20of%20gamin. Accessed 15 Mar 2024

Prohibition of Advertisement and Regulation of Trade and Commerce, Production, Supply and Distribution Act 2003, Cigarette and Other Tobacco Products, ACT, 2003. https://www.indiacode.nic.in/bitstream/123456789/15366/1/the_cigarettes_and_other_tobacco_products_%28prohibition.pdf. Accessed 15 Mar 2024

Rahim A (2022, September) Surrogate Advertisement in India – the Legal Loophole. Int Res J Modernization Eng Technol Sci 04(09). https://www.irjmets.com/uploadedfiles/paper/issue_9_september_2022/29776/final/fin_irjmets1662843965.pdf. Accessed 15 Mar 2025

Singh C (2015) Evaluation of surrogate advertising and its legal measures with special reference to India. Manage Insight:103–107

Supra Note 6, 9,12, 24

Swasticharan L, Arora M, Sinha P, Ray CS, Munish VG, Shrivastav R (2022) Report on tobacco control in India 2022 (Vol 2). Ministry of Health and Family Welfare. Government of India (PDF) Report on Tobacco Control in India 2022. https://www.researchgate.net/publication/380789959_Report_on_Tobacco_Control_in_India_2022. Accessed 15 Mar 2024

Tanzila S(2023) Betting firms take the Surrogate Advertising route: A fair play? https://www.exchange4media.com/advertising-news/the-surrogate-route-can-government-really-stop-betting-companies-from-advertising-126575.html. Accessed 15 Mar 2024

Wells HG (2009). Quoted in The Oxford dictionary of quotations, 7th edn, Oxford University Press, Oxford

Ankur Taya is a law student and currently Junior Research Fellow (JRF) in the Department of Laws, Panjab University, Chandigarh. He has practised as Advocate in District and Sessions Courts, Karnal, Haryana, and the Punjab and Haryana High Court, Chandigarh. Having completed B.A. LL.B. Hons. and Master of Laws from Panjab University, Chandigarh, he has been actively participating in various conferences, seminars, and youth festivals held at college/university levels and has passionately represented cricket teams at the college level and is currently a member of the District Bar Association Karnal Cricket team. Headed the contingent and participated in Harvard National Model United Nations held at Harvard University, Cambridge, MA, USA. Currently, he is researching "Laws relating to Betting and Gambling".

Prof. (Dr.) Aman Amrit Cheema is a distinguished academic and legal scholar, currently serving as a Member (Faculty of Law), Panjab University, Chandigarh. With an illustrious career in legal education and research, she has made significant contributions to the field of law, particularly in Sports Law and Criminal Law.

Dr. Cheema has served as the Director, Panjab University Regional Centre, Ludhiana, where she played a pivotal role in advancing legal education and fostering academic excellence. She is an accomplished author, having written a book on 'Sports and Law' and 'Narcotics and Law'. Her scholarly output includes 54 articles in refereed peer-reviewed journals and has also contributed numerous chapters and write-ups to various books, showcasing her expertise in diverse legal domains.

As a sought-after academic, Dr. Cheema has delivered 35 invited lectures, sharing her insights at prestigious platforms. She has also served as a Visiting Faculty at the Indian Institute of Management (IIM), Rohtak, Haryana, where she enriched the academic environment with her expertise. Recognized for her contributions to criminal law, she holds the distinguished title of Honorary Professor of Criminal Law, International Institute of Justice and Police Sciences, Bengaluru, Karnataka. She is also Member (Advisory Board), The Communications, Centre for Distance and Online Education, University of Kashmir, Srinagar.

Dr. Cheema is the Editor-in-Chief of the Journal of Law and the Student Law Journal of Panjab University Regional Centre, Ludhiana, where she guides and shapes legal discourse through editorial leadership. Her multifaceted contributions make her a prominent figure in legal academia and research.

Chapter 6
Constitutional Welfare Economics: A Study of Classical and Keynesian Economic Ideas in Indian Constitutional Economic Philosophy

Rohtash and Devinder Singh

Abstract Constitutional Economics is a set of legal, institutional, and constitutional decision-making for allocating and distributing societal resources, revealing the preferences for social choices. It determines the nature and national strategies for development, shapes public policies, legislates on subject matters, and executes and institutionalizes the legislative intent. Constitutional welfare economics is that part of constitutional economics that sets the growth prospects, idealized development philosophies and ideas, and characterizes both the short-term and long-term public policies. Constitutional welfare economics is mainly a feature of democratic constitutions and open societies. Economic constitutionalism is a recent reactionary and protective phenomenon against global capitalism in the national, regional, and local interests of the people. The Constitution of India is a social document that embodies welfare ideas from both classical and Keynesian economics. This research paper aims to explore the Constitutional economic framework and welfare approaches that facilitate and legislate economic liberty, specifically the rights to carry on business, profession, and occupation, the right to property, and freedom of association, in order to promote economic welfare and liberty.

Keywords Constitutional economics · Economic constitutionalism · Constitutional welfare economics

The chapters in this book are presented in the "International Conference on Legal and Economic Synergy for an Equitable Society, organised by the University Institute of Laws at Panjab University Regional Centre, Ludhiana, India, on February 1st, 2024.

Rohtash (✉) · D. Singh
Department of Laws, Panjab University, Chandigarh, India
e-mail: rohtash.rohi22@gmail.com; devinder@pu.ac.in

A. Virk et al. (eds.), *Law, Economy, and Society: Insights into Legal and Economic Issues in India*, Accounting, Finance, Sustainability, Governance & Fraud: Theory and Application, https://doi.org/10.1007/978-981-95-3361-9_6

6.1 Introduction

Constitutional welfare is related to the governance that determines principles and policies, the nature of the economic development, limits the markets within the social choice and welfare, and bounds the State to act to achieve the Constitutional goals and materialize the constitutional ideas it carries. Constitutional economics, as a normative science, develops governance-based policies and principles that are subject to moral and value judgments. Therefore, it emphasizes the need to develop a value judgment system to evaluate governance. The constitutional welfare economic philosophy of the Indian Constitution is both egalitarian and utilitarian. The constitutional norms are the source of logical reasoning behind the constitutional economic governance. The constitutional rules and arrangements also aid in interpreting the constitutional provisions to construe. The constitutional conventions and traditions, to some extent, aid in the economic interpretation of the legislative intent. The understanding of constitutional welfare economics depends upon the three different approaches: first, the predictive analysis, second, functional interpretation, and third, the normative analysis (Backhaus 2001). The economic functions of a constitution are that it ensures, through the legislation or public policy, that the allocative and distributive efficiency, both at the community level and the individual level, trickle down the benefits of economic growth, participation of all economic agents in the market. The economic models followed by countries must be within the constitutional governance. The development strategy and model followed by early Indian economic thinkers and planners were influenced by both the economic ideas of the Classical economists and Keynesian economics. The development strategy during the planning period (1951–1991) was influenced to a greater extent by Classical ideas, while both Classical and Keynesian economic ideas influenced the reform period (1991 onwards). The Indian economic planning is centered on the Preamble and the Directive Principles of State Policy. Economic constitutionalism is an economic and political process that transforms political democracy and freedom into economic democracy by creating equal opportunities for all, irrespective of class and caste. Economic constitutionalism is the basis of social democracy. The survival of the Constitution and political democracy depends upon social democracy. Social democracy is only possible when we are able to transform the socio-economic conditions of the society. If a constitutional philosophy and ideas allow political democracy to grow into social and economic democracy, then there would be no wars, no resistance, no strife, and no totalitarian form of Government.

Constitutional welfare economics is the research area that analyzes the preferences and choice of rules by using economic principles and rationality. The growth of constitutional economics was caused by the expansion of globalisation after the 1980s. Both classical and Keynesian approaches have failed to solve the economic problems in the highly globalised world. Constitutional economics provides the guiding principles for the formulation of fiscal and monetary policy. The welfare constitutional economics force states to take the fiscal deficit, public debt, manage inflationary pressure, and use public resources in the best interests of society. The constitutional provisions regarding the fiscal rules and public expenditure are the stabilizing tools to achieve fiscal stability and balance budgets. The 91st amendment in the Constitution of India is the finest example of how the rational macroeconomic management to achieve macroeconomic stability is designed in constitutional economics.[1] The 91st Constitutional Amendment Act, 2003 completed by the Parliament to achieve the Preambular ideas stated in the Fiscal Responsibility and Budget Management Act, 2003 stated as, "An Act to provide for the responsibility of the Central Government to ensure intergenerational equity in fiscal management and long-term macro-economic stability by removing fiscal impediments in the effective conduct of monetary policy and prudential debt management consistent with fiscal sustainability through limits on the Central Government borrowings, debt and deficits, greater transparency in fiscal operations of the Central Government and conducting fiscal policy in a medium-term framework and for matters connected therewith or incidental thereto."[2] The parliamentary control over finance in the Indian Constitutional framework demonstrates that the discretionary role of the state, particularly that of the executive, is limited by the Constitution, which is a primary reason for the growth of the Constitutional economy in India. Article 112 carries the provision of the annual financial statement in the parliament, the expenditure shall not be drawn from the Consolidated Fund of India without passing the Finance Bill or money. There is a distinction between the expenditure charged upon and the expenditure made from the Consolidated Fund of India. Therefore, the constitutional provision is unambiguous that the government has limited discretionary power over finance.[3] The clear constitutional provisions on the Money Bill and the Finance Bill limit the power of the government with respect to public expenditure.[4] (Herbert 1938) It is the classical work on the welfare character of the US Constitution. It identifies the source of welfare legislation in the United States Constitution. This research work pointed out that civil society and welfare in the constitution are by-products of such civilisation. The basic Constitutional principles and philosophy are responsible for the socio-economic legislation. The welfare character of the Constitution is limited in the world. The US Constitution vested power in the Congress to do in the general welfare of the United States.[5] The spending clause is entirely in the domain of Congress, where courts are only to evaluate the welfare within the scope of the powers vested in Congress, not beyond. The discretion on how to spend public money on the welfare of the United Nations is vested or lodged in the Congress rather than the courts, through legislation or public policy.[6] The

---

[1] Article 75(1A) instituted by the 91st constitutional amendment Act, 2003 stated as 'The total number of Ministers, including the Prime Minister, in the Council of Ministers shall not exceed fifteen per cent. of the total number of members of the House of the People'.

[2] Preamble, Fiscal Responsibility and Budget Management Act, 2003.

[3] Article 112 of the Constitution of India stated as, *Annual Financial Statements.*

[4] Articles 109, 110 and 117 of the Constitution of India, related to the Money and Finance bills and procedure thereof.

[5] Article 1, Section 8, Clause 1, The Constitution of United Nation.

[6] Lyng v. Int'l Union, 485 U.S. 360, 373 (1988).

scope of the judicial review is irrelevant to courts to analyze whether the public spending is excessive, unwise or wasteful to the general welfare.[7] (Barber 2003) This book aims to examine the various welfare dimensions of the US Constitution. It explains that state-facilitated welfare is the result of policy experiments by legislation and the philosophical framework of the US Constitution. The author in Chapter III explains that constitutionalism may be positive and negative in the context of welfare. Positive constitutionalism in the book is a similar concept to economic constitutionalism in this research. This research work questions whether the Constitutional Institutions are empowered sufficiently to empower society. The establishment of the welfare character through the welfare clause in the Constitution is a rational act as a promoter of economic constitutionalism (Lowi 1986). The research paper hypothesized and proved that if an anti-statist State acted on the Constitutional ethics, it could transform itself into a welfare state without any social revolution or reforms. It explains that the trilogy of rights, entitlement, and privilege is the basis of welfare in the democratic Constitution. Economic constitutionalism is possible under the conditions of rule of law, liberalism, and self-reliant (Lina Takeshi and Matsubayashi Tetsuya 2010). Constitutional designs constrain the interests of the elites in society and emphasize the discourse of the majority and citizens' rights to receive the welfare services and on the basis of the priority of the State. The foundational philosophies and political principles support and shape the welfare policies. The constitutional provisions regarding social welfare exert a strong influence on the socio-economic legislature, public spending, and welfare policies. A more liberal constitution strengthens economic constitutionalism. Constitutional welfare economics is the settlement of competing interests of the liberal elites and conservative elites. The liberal elites tend to support the principles of egalitarianism, while the conservative elite are in favour of the principles of individualism (Gerber 2001). Economic constitutionalism became a fundamental principle of the social market economy and a blueprint for economic prosperity that is economically, socially, and politically just. The economic constitutional governance is flexible it designs and shapes the economic policies according to the objectives set by the legislation in order to achieve the constitutional goals through the transaction economy or by the centrally commanded one. The constitutional economic governance combats the global competition and protects and promotes the domestic economic agents against such a new competitive economic order. The economic constitution has the potential value to influence the international economic orders and international institutions and their relationship with states and societies (Skordas et al. 2023). The concept of economic constitutionalism is both a normative framework and a descriptive referring to principles, rules and practices to influence the State, public and private economic networks, and transnational institutions. It structures manages economic problems, deepens and strengthens economic relations and expands the scope of the public goods and enlarges economic choices and economic freedom. The vision of economic constitutionalism is all about the integration of the

[7] Mathews v. De Castro, 429 U.S. 181, 185 (1976).

social choice theory through the constitutional adjudication and constitutional inclusive procedures, which contribute to producing democratic decision-making and social welfare. The European Union's economic constitutionalism is the best example that economic constitutionalism is a movement in the post-2008 global crisis (Mueller 1973). The constitutional economic is related to two types of stages: the first is the choice making, and the second is the decision making. The collective choices by the methods of the democratic process depend upon the two stages, the first is Constitutional and the second is Parliamentary. The proper procedure for allocating and distributing societal resources among the different sectors of society and among individuals is the core subject matter of constitutional welfare economics. The draft of the constitution is an example of collective choice-making, while the elected representatives and the form of government are examples of collective decision-making.

Hypothesis The Constitution of India is conceptually influenced by classical and Keynesian economic ideas with respect to economic development and to maintain the macroeconomic stability framework.

Classical economic orientation towards the advancement of the common interest as found in the Constitution of India in the provisions of Part IV, the Directive Principles of the State policies, and the preamble it carries. The classical economics emphasis on the role of the political institutions, the constitution of India, also depends upon the constitutional agencies for the purpose of social welfare. The principles of egalitarianism and utilitarianism are found in the basic philosophy of the Constitution of India. Egalitarianism focuses on each and every individual in the society, irrespective of the social, political, or economic status. The egalitarian philosophy found in the Part III the Fundamental Rights, Article 14 *the equality before law and equal protection of law*, Article 15 and 16 *prohibition of discrimination on the basis of the religion, race, caste, sex and place of birth*, Article 19 gives space to both utilitarian and egalitarian philosophies *the right to trade, business, occupation and profession* is based on the sound philosophy of the regulated market and Article 21 *the right to life and personal liberty*. The *invisible hand* ideas of Adam Smith are found in certain freedom clauses of Article 19 of the Constitution of India. Economic liberty is best secured by certain freedoms guaranteed in Part III in the form of Fundamental Rights and in Part IV as the Directive principles of State Policy. The right to private property under Article 300A[8] The right to property is the most striking economic right in the constitution, it is the source of the economic freedom and liberty based on the classical principles of individual utilitarianism (Jeremy 1781). The free flow of property and the security of the property guarantee that individuals can enjoy the fruits of their own labour, which further promotes the *division of labour* in society. Adam Smith argued that the tolerable administration of justice to secure and strengthen the property rights attracts and allows the private investment

[8] Earlier it was fundamental right under Article 19(1)(f). The forty fourth amendment place this right by inserting a new Article 300A, it is mere a Constitutional right now.

and exchange, that is one of the sources of economic growth in the short run and economic progress in the long run (Irwin 2014). Adam Smith advocated the theory of natural liberty i.e. the organising the economic activities and free enterprises. The same ideas are reflected in Part IV the Directive Principles of State Policy and XIII the Trade, Commerce and intercourse within the territory of India in the Constitution. Article 38 of the Constitution is the source of economic constitutionalism in India. The said Article stated in very strong language that it shall be the duty of the State to promote the welfare of the people by securing and protecting the social order in which economic, social and political justice is promoted in the national life of the country. The economic constitutionalism is only possible when economic governance is able to stop perpetuating the present exploiting social order. Articles 42, 43 and 43A are the source of the labour economy in India. According to classical economics the labour is an important factor of production. The labour legislation should be focused on the labour welfare and production relations in society. The Constitution provisions related to labour are influenced by the Marxian ideas and trade union movements during the period of industrial revolution Articles 38, 42, 43 and 43A carry the foundational principles of labour welfare in India. Article 39(d) is the striking constitutional provision that states that the State must follow the policy of equal pay for equal work. The same Article 39(e) stated that the State must ensure the health and strength of the workers. Article 42 stated the provision of just and humane conditions of work and maternity relief. The core idea behind Article 43 is that suitable economic legislation must ensure the level of wages. The concept of living wage and minimum wage is a constitutional guarantee under Article 43. The participation of the workers in the management of the industries is based on the Marxian philosophy that workers manage industries according to the social needs of society, but not for the greed of the profit of the capitalistic class. The preambular ideas in the Constitution of India are to establish an egalitarian society. The social and economic democracy that the Constitution provides is an instrument of instruction to meet the changing needs of the succeeding generations. Economic constitutionalism is a social and economic engineer by reducing income inequalities and removing social disabilities, and by creating a just social order. The taxation Constitutional provisions are influenced by the ideas of John Maynard Keynes. Article 265 is the source of the legislative framework on the taxation system in the country and makes taxation a legally compulsory payment levied by the authority of law. J.M. Keynes was a supporter of State intervention in economic activities. The well-regulated and controlled public economic and public finance in the Constitution of India limits the power of the state to act in a discretionary. The taxation provisions in the Constitution are the constitutional obligations of both the legislation and the executive. ".... Constitution to be a contract between the State and the people. This contract has been given for drafting to the arbitration of the representatives of the people. We should therefore not be guided, biased, or prejudiced by the administrative difficulties...but we should take notice of the difficulties of the citizens at large. The Constitution is a contract between the Citizen and the State; the main terms are that the citizen shall pay such and such taxes whenever they are required by law to do so. This is the biggest liability which the citizens agree to take on themselves. The state, through this Constitution, takes over the responsibility of maintaining peace and enhancing the prosperity of the people.... We see that the State is not actually rendering any free service except that we are giving our citizens a psychological satisfaction to enjoy in their belief that they are a free people. They do not know the value of freedom. State justifies taxation on the plea that it defends the borders against wars. Wars never come alone, and when a war comes, it brings an extra tax along.[9] The Constitution of India is originally a welfare economic document that guides the State in matters of tax legislation. The State shall not impose a higher level of taxation without the statement of objects and reasons. Article 265 clearly stated that tax is a legal compulsory payment. The Parliament cannot delegate the legislative power over taxation. The Constitution of India leaves the adoption of any kind of taxation system in the light of the preambular ideas and the Directive Principles of State Policy that the taxation system should be based on the social maximum welfare. The close examination of the Indian taxation system suggests that it is influenced by the ideas of Arthur Pigou and Pareto Optimality. The Constitution is the source of both the direct,[10] income tax[11] and indirect tax legislation.[12] The combined operation of Article 110 (Money Bill), Article 112 (Annual Financial Statement, Article 265 (tax as a legal compulsory payment), Article 269A and 297A (GST), Article 280 (Finance Commission), and Entry 82 of the Union list in the Schedule VII is the source of the Public Economy in India.

Hypothesis Analysis The hypothesis was proved by examining the working of the Constitutional governance in India in the context of the macroeconomic stability framework. Part III, the Fundamental Rights, and Part IV, the Directive Principles of State Policy of the Constitution of India, reflect the philosophy of both the Classical and Keynesian economic ideas. The public economics, the monetary economics, and the development economics are guided by the Constitutional welfare economics. The Law, development, and planning in the story of Indian economic development is the finest example of both Constitutional economics and economic constitutionalism.

6.2 Conclusion

The classical political economic ideas are reflected in the Constitution of India to maintain macroeconomic stability and the patterns adopted for economic development in the history of Indian economic planning. The development and economic growth story of the Indian economy shows that the patterns and strategy followed for economic development are influenced by the classical economic ideas in the

[9] The Speech by Prof. Shibban Lal Saxena, The Constituent Assembly Debates on 16 October, 1949.

[10] The Article 265 of the Constitution of India.

[11] Entry 82 of Union List, Schedule VII of the Constitution of India.

[12] The Article 296A of the Constitution of India.

early days of independence. When the country needed a strong role of the state to build a strong base, i.e., infrastructure, all the decision making in all the coordinating tasks that is what to produce, how to produce and for whom to produce is rest on the State to the large extent, the private sector played a minor part in the planning period. The constitutional economic goals, which are carried by the preamble and Directive Principles of State Policy, were achieved by formulating the economic planning influenced by the ideas of Keynesian and Classical welfare economics in the planning period. The Planning period is dominated by the central decision-making and characterised by the commanding heights period in the Indian economy. The planning period gave more importance to the public sector rather than the market, while the reform period reduced the considerable role of the State and left the economic decision-making to the market. The reform period is influenced by the classical economic ideas and more focused economic liberation, which gave importance to the private sector too, and globalised the domestic economy with the rest of the world. The idea of distributive justice is realised and materialised under the influence of preambular ideas and the Constitutional provisions of Part IV, the Directive Principles of State Policy, with the help of the State. In the reform period, it was realised that in the distribution, the market is more efficient than the State.

The constitution of India reflects the philosophy of egalitarian, utilitarian schools and the Rawlsian concept of justice. Economic constitutionalism is the essence of the living constitution; the judiciary plays an important role by applying the egalitarian and utilitarian principles in the application and interpretation of the Constitutional provisions. Utilitarian principles found in both Part III and Part IV of the Constitution of India, the Fundamental Rights are available to both the individual level and community level, while the Directive Principles of State Policy are only available at the community level. Part III of the Constitution reflects both utilitarianism and individual utilitarianism. The preambular ideas of justice are a multi-dimensional concept as it includes the economic justice, 'creating equal economic opportunities for all', which reflects the individual utilitarianism. The constitution adopted a mixed economy, and the planned development has become a constitutional scheme to realise an egalitarian social order. The second Five-year plan envisaged that "The patron of development and the structure of the socio-economic relation should be so planned that they result in appreciable increase in income and employment but also in greater equality in income and wealth." The Directives of the state policy have delineated a broad spectrum of socio-economic justice to all people. "The socialistic patron of society is a more comprehensive expression of the approach. Economic policy and constitutional changes have to be planned in a manner that would ensure economic advance along with democratic and egalitarian lines." Economic justice means abolishing such economic conditions that remove the inequality of economic value between men, the concentration of wealth and means of production in the hands of a few, and are detrimental to the vast. Law, therefore, must seek to serve as a flexible instrument of socioeconomic adjustment to bring about peaceful socioeconomic revolution under the rule of law. Economic constitutionalism is not possible only if the mere Constitutional provisions carry the principles behind the economic development, but the constitutional, social, and political institutions also make economic constitutionalism possible in India. The constitutional welfare economics in India is to be needed in accordance with the ideas, hopes, and aspirations imbibed in the Constitution of India by the founding fathers of the Constitution. The Gandhian principles in the Directive Principles of the State Policy in Part IV of the Constitution are the guiding principles for the inclusive growth that brings the underprivileged section into the mainstream of Society. The trade and commerce provisions in Part XIII of the Constitution of India are also influenced by the philosophy of both the Classicals and Keynes. As the Indian economy is witnessing more globalising phenomena, the ideas of Classical and Keynesian will be more effective in achieving macroeconomic stability. Law, Development, and planning in India are influenced by Keynesian ideas during the early days of independence, while Classical ideas dominated after the reforms period (after the Economic Reforms 1991). The economic freedom in the Constitution of India is reflected in the form of the right to property as a Constitutional right, the right to carry on business, profession, occupation, and trade as a Fundamental Right, and full freedom to contract. The commercial and industrial philosophy in the Constitution in Parts III and IV is clear evidence of the Classical and Keynesian economic ideas. As far as the Federal finances are concerned, the Finance Commission under Article 280 is established as a Constitutional agency where the States in the Union are to a large extent dependent upon the Union. Similarly, monetary policy controls all the interests of the Union and the States, having no control over the elements of the monetary policy. The commercial interests of the States placed and converted under the fiscal policy of the Union. The economic potentialities of the States have been undermined in the economic federal design of the Constitution. Similarly, as the Indian economy integrated with the rest of the world, the commercial interests of the States were also undermined. Therefore, in the reformed period, it was experienced that States should be needed more economic and commercial freedom in the Constitution. There is an urgent need of economic reconciliation in the Constitutional scheme about the economic unity of the Union and of the States. It is respectfully submitted that all the revenue and expenditure related to the national building aspects, institutions, and Constitutional agencies should be assigned to the Union, while all other subject matters and expenditure and revenue related thereto should be left to the States by the Finance Commission. As far as the direct and indirect taxes are concerned, a certain percentage of income tax should be handed over to the States, subject to certain limitations. Article 112 should be amended and reconsidered in accordance with the expenditure charged upon and the expenditure made from. The Constitutional scheme of parliamentary control over the finances of the Union should be rethought. It is also submitted that the States having customary balanced budgets should be given more economic freedom in the Union budget and by the Finance Commission, while the States are facing deficit budgets due to geographical, natural calamities, and unavoidable circumstances. The budgetary credit States should receive benefits equivalent to their credits, and debit States should receive extra benefits from the Union. The customary rights of the States should be financially supported in the constitutional scheme. The States that historically create financial obligations and difficulties for the Union should be categorised into the Constitutional temporary design. The economic unification should be guided by the principles of the federal and democratic polity. The Reserve Bank of India should be transformed from a Statutory body to a Constitutional body in the Constitution. As far as the labour welfare is concerned, the vested interests of the capitalists are the overwhelming majority, as the labour welfare legislation is under Part IV of the Constitution. The provisions in Part IV of the Constitution depend upon the economic capacity, desire, vision, and political environment of the country. Therefore, the progressive and welfare legislation seems difficult in the present Constitutional Scheme. The welfare of the labour according to the Classicals depends more on the market forces than on the State, while Keynesian philosophy stressed the role of institutions in this regard. The Trade Unions play an important role in the labour welfare legislation. In classical economics, the trade unions as a unit influenced both the cost of production and productivity. In the present state of globalisation, the Constitution should be stringent on the economic, commercial, and business principles without unnecessary interference and control.

References

Article 1, Section 8, Clause 1, The Constitution of the United Nations
Article 112 of the Constitution of India states, Annual Financial Statements. https://indiankanoon.org/doc/280240/. Accessed 18 Mar 2024
Article 75(1A) instituted by the 91st constitutional amendment Act, 2003 stated as 'The total number of Ministers, including the Prime Minister, in the Council of Ministers shall not exceed fifteen per cent of the total number of members of the House of the People
Backhaus JG (2001) Economic principles of the constitution: an economic analysis of constitutional law. The Independent Institute Working Paper 40(7), 2–24
Barber A (2003) Welfare and the constitution. Princeton University Press, pp 124–243
Earlier, it was a fundamental right under Article 19(1)(f). The forty-fourth amendment placed this right by inserting a new Article 300A; it is merely a Constitutional right now
Entry 82 of the Union List, Schedule VII of the Constitution of India
Gerber DJ (2001) Economic constitutionalism and the challenge of globalization: the enemy is gone? Long live the enemy: comment. J Institut Theor Econ (JITE) / Zeitschrift Für Die Gesamte Staatswissenschaft 157(1):14–22. http://www.jstor.org/stable/40752244
Herbert HJ (1938) The general welfare clauses in the constitution of the United States, 7 FORDHAM L. REV 390. Available at: https://ir.lawnet.fordham.edu/flr/vol7/iss3/5, https://ir.lawnet.fordham.edu/cgi/viewcontent.cgi?article=1161&context=flr
Irwin DA (2014) Adam Smith's "Tolerable Administration of Justice" and the Wealth of Nations, pp 56–78. National Bureau of Economic Research. Working Paper, No 20636
Iida T, Matsubayashi T (2010) Constitutions and public support for welfare policies. Soc Sci Q 91(1):42–62. http://www.jstor.org/stable/42956522
Jeremy B (1781) An introduction to the principles of morals and legislation. Prometheus Books, pp 124–129
Lyng v. Int'l Union, 485 US 360, 373 (1988)
Lowi TJ (1986) The welfare state: ethical foundations and constitutional remedies. Polit Sci Q 101(2):197–220. https://doi.org/10.2307/2151388. Published: 15 March 1986
Mathews v. De Castro, 429 US 181, 185 (1976)

Mueller DC (1973) Constitutional democracy and social welfare. Oxford J 87(1):69–76
Preamble, Fiscal Responsibility and Budget Management Act, 2003., https://indiankanoon.org/doc/58621322/. Accessed 18 Mar 2024
Skordas A, Halmai G, Mardikian L (2023) Economic constitutionalism in a turbulent world. Edward Elgar Publishing, pp 167–271
The Speech by Prof. Shibban Lal Saxena, The Constituent Assembly Debates on 16 October 1949

Rohtash is Assistant Professor, Department of Laws, Punjab University, Chandigarh; Ph.D. in Law on the topic "Emerging Trends of Distributive Justice in Indian Context with special reference to Economic Legislation: An Analysis"; and LL.B., LL.M., and Master's in Economics from Panjab University, Chandigarh. The teaching areas are constitutional law, criminal law, election law, law and economics, and law and public policy. The research areas are constitutional law, law and economics, and election laws. Published more than ten publications in national and international journals. Delivered special lectures on the Constitutional laws, interpretation of Statutes as a resource person. Awarded the "Best Discussant" in the "4th GNLU Workshop on Economic Analysis of Crime and Constitutional Law" organized by the Centre for Law and Economics, Gujarat National Law University on 2nd August and from 8th to 10th August 2023. Delivered special lecture on Interpretation of Statutes on 23rd July 2024 at the office of the National Academy of Customs, Indirect Taxes and Narcotics, Zonal Campus, Central Revenue Building, Plot No. 19, Sector-17-C, Chandigarh, Central Board of Excise and Customs (CBEC), Department of Revenue, Ministry of Finance, Government of India. Delivered Keynote Lecture on Drug Abuse, Phone Addiction and Road Accidents: Youths in Public Health Crisis at NSS Seven Days/Nights Special Camp (10th March–16th March 2023) in D.A.V College, Sector-10, Chandigarh, in collaboration with the Ministry of Youth Affairs and Sports, Government of India, under the G-20 awareness campaign "Empowering Youth for Sustainable Development" on 10th March 2023. Delivered special lecture on the topic "Mainstreaming of SCs, STs and OBCs in the Constitution of India", organized by the Centre for the Study of Social Exclusion and Inclusive Policy, Panjab University, Chandigarh, held on 25th May 2022. The major publications are Constitutional Economics: A survey of Indian Constitution, "Academic Integrity, Research Ethics and Issue of Plagiarism", "Racial Justice in USA: A Critique", "Egalitarian and Utilitarian Philosophy in Indian Constitution", and "Distributive Justice and Global Public Healthcare system with Special Refence to COVID-19 vaccination Programme".

Prof. Devinder Singh, Vice Chancellor of Dr. B.R. Ambedkar National Law University (DBRANLU), Sonepat, researches and teaches in the areas of constitutional law, administrative law, and election law. Published research papers in nationally and internationally reputable journals. Life Membership of the Institute of Constitutional and Parliamentary Studies, New Delhi, Life Membership of the Institute of Public Administration, New Delhi, Voting Member, Global Alliance for Justice Education, Life Membership, Indian Law Institute, New Delhi, Membership, International Association of Constitutional Law || l'Association Internationale de Droit Constitutionnel, Law and Society Association, Chicago, USA, till 2025. Authored seven books such as *An Introduction to Administrative Law*, Allahabad Law Agency, Faridabad, 95759-74-8, 2023, *Law, Religion and Secularism, Historical, Constitutional and International Perspectives*, SLM Publisher, Patiala, ISBN 978-93-91083-41-0 (2022), *Human Rights: Women and Law*, Allahabad Law Agency, ISBN 81-89530-68-2 (3rd Edition 2024 and First Edition in 2005), *Administrative Law*, Allahabad Law Agency, ISBN 978-93-80-231-00-6. (3rd Edition 2024 and First Edition 2008) (Sole Author), and *50 Years of Basic Structure Doctrine: The Way Ahead*, SLM Publisher, Patiala, 978-93-91083-87-8, 2024. Presented paper titled "Imperial resonance: The colonial legacy's imprint on constitutional rule of law" In IACL Roundtable, Rome: Current Challenges to Constitutional University of Sapienza, Rome on 3rd and 4th October 2024. "Students' Perspectives and Reflections on Clinical, held in Chiang Mai, Thailand, from 26 to 28 May 2023. Presented a paper titled "Role of Legal Aid Clinic in Socio-legal mobilization: Case

Study of Punjab" in Worldwide 2022 GAJE/IJCLE/SAULCA Conference held by the Global Justice Education, the International Journal of Clinical Legal Education, and the South African Association in Collaboration with the Stellenbosch University of Stellenbosch, from 11 to 15 December National (UGC-Care), International Journals (Scopus-Indexed), Lectures, Panelist, and Chairmanships in International, Seminar, Conference, and Symposiums: 148 Courses/ Refresher Courses organized by HRDC evaluation programmes/faculty development programmes—Academic/Administrative Responsibilities in Public and Private Universities Seminars/Lectures/ Workshops/Competitions organized in capacity of the Chairperson, Panjab University, Chandigarh.

Part III
Technological Advancements and Financial Systems

Chapter 7
Laser Levelling Technology for Sustainable Irrigation in Sangrur: An Economic Analysis

Priya

Abstract Sustainable agriculture provides profitability while balancing the requirements of the present and future generations. This study analyses the economic impact of Laser Levelling Technology (LLT) upon water productivity, associated cost, and crop yield in the case of the wheat crop from the Sangrur district. It also attempts to understand the socio-economic determinants of the adoption of farm households. The analysis is based on secondary and primary data, and also a comparative study. The primary data comprised a limited sample of 56 farm households, with 36 adopting this technology and 20 serving as the control group. The analysis of data establishes that the farmers who had adopted this technique realized better water productivity, saved on the front of cost, and realized better returns. Moreover, highly educated and large farm households have adopted more of the latest irrigation techniques. The study, overall, argued for the adoption of this technology with apparent benefits. Therefore, the government needs to adopt policies and incentives aimed at water use to ensure the sustainable development of the agricultural sector.

Keywords Sustainable agriculture · Irrigation water · Economic analysis · Punjab · India

The chapters in this book are presented in the "International Conference on Legal and Economic Synergy for an Equitable Society, organised by the University Institute of Laws at Panjab University Regional Centre, Ludhiana, India on February 1st, 2024".

Priya (✉)
Silver Jubilee Hostel Panjabi University, Patiala, Punjab, India
e-mail: manpriya193@gmail.com

A. Virk et al. (eds.), *Law, Economy, and Society: Insights into Legal and Economic Issues in India*, Accounting, Finance, Sustainability, Governance & Fraud: Theory and Application, https://doi.org/10.1007/978-981-95-3361-9_7

7.1 Introduction

Sustainable agriculture balances the needs of current and future generations while simultaneously ensuring profitability, environmental, social, and economic justice. That is why sustainability practically works through netting and nesting the economic, social, and ecological dimensions by suitable technological interventions. Agricultural production structures and associated food systems have been facing growing competition with alternative uses of limited natural resources. The availability of resources has been laid low by the prolonged grip of ongoing unsustainable practices. Water remains the primary natural resource for agricultural production systems worldwide. Many communities around the world face water scarcity because water is not sufficient, both in terms of desired quantity and quality, which necessitates its rational use and conservation.

India is an agricultural economy, and the country's overall economic performance is highly dependent on the performance of the agricultural sector. India's economic growth rate is influenced by everything that happens to its agriculture. Given the great need for food security in India and its rapidly growing population, this is also a national priority. India's food security is threatened by several issues, including water shortages that require urgent attention. Solutions aimed at addressing food security issues, such as improved varieties and improved management practices, cannot be fruitful without water.

Water development and conservation have become a major concern in India, as some parts of India are suffering from severe water shortages. Water problems are worsened by increased consumption in the summer. With increasing demographic pressure, the development of appropriate institutional arrangements surrounding water is crucial to the conservation and sustainability of the country's water resources. India has taken several steps to save water in agriculture by producing alternative crop mixtures to ensure optimal water use. The Union Minister of Jal Shakti recently said that the total groundwater available, 6%, is used for household use, and an additional 5% is used for industrial use. The remaining 89% is used for agriculture. "He said, out of 178.7 million rural households, only 32.7 million (18%) received tap water. And by 2050, India will become a global hotspot for "water insecurity."" It is further held that the focus should shift from land productivity to irrigation water productivity so that farmers can work on saving water (Bakshi 2019). Keeping water management in view, the study is divided into five sections. The following section describes the water issues in Punjab agriculture. Section 7.3 provides in detail the data and methodology. The results are presented in Sect. 7.4 with a detailed focus on cost, productivity, and returns. The final section concludes and provides some policy recommendations.

7.2 Water Issues in Punjab Agriculture

Water is what humans need. People working in different fields need water for different purposes, such as agriculture, industry, and home. Due to the inadequate availability of water of the required quantity and quality, water shortages are faced in many parts of the world, requiring careful use and conservation of water. Water conservation refers to the proper use of water with caution, strategies, and policies to prevent waste, as water is in great demand for the current population and future generations.

Irrigation development plays an important role in improving farmers' incomes through improving agricultural productivity, cultivation intensity, employment opportunities, and wage rates. The supply of irrigation water can be improved by adding new reservoirs and by adjusting for irrigation water loss. One of the options available to regulate field loss and increase farm water productivity is to replace traditional irrigation systems with modern irrigation systems.

Punjab's agricultural system, based on the mono-cropping pattern of rotation of paddy and wheat, requires intensive use of water for irrigation throughout the year. The Punjab state contributes 36.8% of wheat and 27.3% of rice to the country's central pool of food grains (GOP 2016). The future of food production in the state will be seriously threatened by unsustainable groundwater use, which at present is open flooding of fields, and thereby in itself is a case of poor water use and hence saving when viewed in the context of emergent irrigation technologies in many parts of the world. The demand for water in the state has been growing and it remains a major issue as the state is a predominantly agricultural-driven economy. During 2017, it was reported that total water (ground and surface) availability in Punjab was 35,109.97 Million Cubic meters (MCM); of which groundwater constituted 20,698.94 MCM and surface water 14,411.03 MCM. It was further estimated that by 2020, the water demand will rise to 44,227.87 MCM with a water deficit in the state equivalent to 14,715.11 MCM (GOP 2017). Noticeably, it is reported that the groundwater level has dropped in about 85% of the areas where groundwater is fresh and suitable for irrigation. The groundwater table rose in about 15% of brackish and salt-containing areas. In the Central Plain Zone of Punjab, the water table is declining at a rate of 54 cm (0.54 m) per year in some patches during 2004 (Aulakh 2004). The decrease in 2019 was 37 cm (0.37 m) per year (Bharti 2019). The excessive decline in groundwater levels has affected the economic viability of small and marginal landowners.

The fast depletion of the groundwater table has been emerging as a big social and political issue also as it involves the economic future of the state. Therefore, the emergent water distress has prompted the state to adopt irrigation water-saving technologies for the last decade or so. Many attempts are underway with the combined efforts of the state, market, and civil society to encourage water-efficient irrigation practices. Many ways have been suggested and practiced to save irrigation water during agricultural operations, such as using central or side-moving sprinklers, minimizing evaporation, using overhead irrigation methods to avoid surface runoff, deep seepage, and underground drainage, and also other micro-irrigation techniques, mainly drip systems.

7.3 Materials and Methods

Keeping in view the precarious subsoil water position of the state, the present study focuses upon the water-saving practices adopted, with a specific focus upon irrigation by scientifically levelling the field. The study analyses the economic impact of the adoption of the Laser Levelling Technology (LLT) upon water productivity, associated cost, and crop yield in the case of the wheat crop. It also attempts to understand the social and economic characteristics of its adoption by farm households. The foremost purpose of this study is to explore the irrigating behavior of farmers by inspecting elements that affect their irrigation choices. It needs to be understood here is that sustainable irrigation water means improving water use efficiency or water productivity by maintaining production level also. The efficient use of irrigation water, alongside its saving, is extraordinarily essential for the sustainability of food security under heavy demographic pressure. The use of irrigation water for agricultural production calls for modern and sustainable research and the correct switch over to water-saving technologies. The study draws its added importance also from the fact that the Government of India has adopted 17 Sustainable Development Goals (SDGs) in 2015, of which managing water resources occupies a central position (GOI 2015). The study is limited to the Central Plain Zone of the Punjab state, where a significant decline in groundwater levels has been observed. With a total area of 34, 42,000 hectares, this zone occupies 43.42% of the state's total area (GOP 2011). This zone consists of eight districts (ENVIS Environment Information System, PSCST, 2020) with the annual average drop in groundwater level (meters per year) from June 1984 to 2017 was as follows: Amritsar (0.30), Tarn Taran (0.39), Kapurthala (0.32), Jalandhar (0.51), Ludhiana (0.33), Fatehgarh Sahib (0.46), Sangrur (0.81), and Patiala (0.73) (Ground Water Resources of Punjab 2020). Importantly, the water table in this zone has dropped by more than 15 meters between June 1984 and June 2016 (Water Resources Department 2018). In addition, depending on groundwater conditions as of March 31, 2017, all 65 blocks in this zone were classified in the 'over-exploited' category (out of 109 over-exploited blocks in the state) and the rest in the 'safe' category (Water Resources Department 2018).

The study has focused on both the secondary and primary data. The secondary data related to groundwater are provided by two government agencies: the Central Ground Water Board, North Western Region, Chandigarh, and the Water Resources and Environment Directorate, Water Resources Department, Punjab, Mohal, with a state-specific publication, namely Ground Water Resources of Punjab State.

Farmers using modern irrigation systems are scattered overall with no definite pattern in the selected zone. The techno-economic nature of the study requires repeated visits and follow-up to generate an adequate dataset. The data with a suitable questionnaire has been collected by incorporating both open-ended and closed-ended questions by the method of personal interviews for one crop season. The data on output has been obtained directly from farmers for The selected crop, wheat, from Sangrur district, has been selected purposely as in this district, the LLT has been adopted extensively. The analysis is based on the comparative study of a limited number of 56 farm households as a sample size, with 36 that adopted this technology and 20 non-adopters as the control group.

The economic analysis has been based on the Benefit-Cost Ratio (BCR). The BCR is estimated by dividing the present value of the estimated benefits by the estimated costs. Similarly, the Water productivity is estimated by using the ratio of yield (kg/acre) to consumed irrigation water (cubic meters/acre). Water usage is estimated by taking into account many factors such as irrigation frequency, well motor output, drilling depth, and flow tube diameter. The information gathered from the primary data is further processed and analyzed to identify the determinants of the adoption of the latest irrigation practices.

The water use impact of the new technique has been examined by using the concept of Water Productivity (WP). It is a term commonly used to describe the relationship between water (input) and agricultural product (output). It is often used to express the effectiveness of irrigation water use and delivery. The water productivity is estimated as the ratio of output (kg/acre) and irrigation water used (cubic meters/acre):

$$WP\left(kg/m^3\right) = \frac{Output\left(kg/acre\right)}{Irrigation\ water\ used\left(m^3/acre\right)}$$

Therefore, to measure groundwater usage by farmers, we used an approximate estimation model following Eyhorn et al. (2005), Srivastava et al. (2009), Watto and Mugera (2015), and Razzaq et al. (2018). The information on the number of irrigations, duration of each irrigation, tube well engine horsepower, the diameter of the suction pipe, and boring depth was collected from farmers to incorporate in the following model to measure the volume of water in liters, which was then converted to cubic meters.

$$\mathrm{Q} = \frac{t \times 129574.1 \times BHP}{\left(d + \left(\left(255.5998 \times BHP^2\right)/\left(d^2 \times D^4\right)\right)\right)}$$

Where Q is the volume of water in liters per acre, t is the total duration of irrigations in hours, d is the boring depth in meters, D is the diameter of the suction pipe in inches, and BHP is the engine power of the tube well in horsepower. This formula gave the extraction of water in liters, which we converted to m^3.

7.4 Results and Discussion

This section provides the results of the study, such as the socio-economic profile of respondents, installation and working costs, estimated values of BCR, and water productivity under LLT and non-LLT users. The analysis of data establishes that the farmers who had adopted this technique realized better water productivity, saved on the front of cost, and realized better returns. Moreover, highly educated and large farm households have adopted more of the latest irrigation techniques.

7.4.1 Production Costs of LLT Versus Non-LLT Users

Table 7.1 shows the average cost per hectare of wheat production for farmers irrigating wheat crops using LLT and non-LLT methods of irrigation. The results show that there was a difference in the cost of various activities between the two types of agricultural systems. Irrigation costs for LLTs are lower than for irrigated farms under the non-LLTs. Labour costs for LLT-irrigated farms were lower than for non-LLT farmers who had used more labour to harvest their crops. Machines accounted for the highest cost of all of the above, but there were also differences in other costs of both LLTs and non-LLT users, and LLT users cost less than those of non-LLT users. The LLT user has incurred comparatively lower costs than those of non-LLT users in each and every cost. The maximum difference was noted in the case of Agro-Chemicals (77.82%) followed by irrigation (78.03%). The use of LLT results in the saving of all types of costs for the farm households. It proved to be cost cost-efficient method of growing wheat with numerous economic benefits.

Similarly, as per Table 7.2, there was a significant difference between yields and revenues for LLT and Non-LLT users. The yield and revenue for LLT users were higher than those of non-LLT users, which indicates that LLTs balanced the supply

Table 7.1 Average cost of cultivation in wheat crop under LLT and non-LLT users, rupee per acre

Costs	LLT Users 1	Per Cent 2	Non-LLT Users 3	Per Cent 4	Cost Gap (Per cent) 5 = 1 to 3
Land preparation	1784	7.94	2045	8.28	87.24
Seeds	2330	10.37	2580	10.45	90.31
Fertilizers	5810	25.85	6014	24.36	96.61
Agro-chemicals	1200	5.34	1542	6.25	77.82
Irrigation[a]	2056	9.15	2635	10.67	78.03
Machinery (Manual Cutting+ Harvester+ Thresher+ Tractor +Other)	7716	34.34	8146	33.00	94.72
Hired labour	1576	7.01	1783	7.22	88.39
Total costs (TC)	22,472	100.00	24,687	100.00	91.03

Note: [a] Includes repair, maintenance, and depreciation of owned irrigation machinery excluding diesel charges, if any. It also includes charges for hired irrigation

Source: Primary Survey

Table 7.2 Crop yield and revenue of LLT and non-LLT users

Category	Yield (qt/acre) (1)	Price (Rs/qt) (2)	By product/acre (Turi Straw) (qt/acre) (3)	Price (By-product) (Rs/qt) (4)	Total Revenue (TR) (Rs/acre) 5 = (1)(2) + (3) (4)
LLT users	23	1975	15	1200	63,425
Non-LLT Users	21	1975	12	1200	55,875

Source: Primary Survey

Table 7.3 Gross return, BCR values and productivity values of LLT and non-LLT users

Category	Gross return (GR) (TR-TC)	BCR values	Water productivity kg/m^3
LLT users	40,953	1.82	0.51
Non-LLT users	31,188	1.26	0.42

Source: Primary Survey

of water to crops than non-LLTs, ultimately resulting in higher wheat yields and therefore revenue also. This makes a strong case for the wider benefits of adoption of the laser levelling technique, as it not only saves water but also gives better yield and reduces cost, making it a better economic choice for the producers. This also indicates a better balance between private profits and social interest in the form of the long-term sustainability of water.

Table 7.3 shows the Benefit-Cost Ratio (BCR) results for LLTs and Non-LLTs farms. It emerged that the gross return of the LLTs users was higher than the gross return of the non-LLTs users. The BCR values for LLTs and Non-LLTs on the wheat crop were greater than 1. The values were 1.82 and 1.26, respectively. These results showed that LLTs and non-LLTs were both economically viable options for wheat harvesting, but the LLTs were comparatively more beneficial. The BCR results proved that a high-efficiency irrigation system (LLT) is an economically more viable and sustainable option, particularly related to the scarce resource that is water.

The water productivity on LLT wheat farms was 0.51 kg/m^{3,} which was much higher than that of non-LLT farmers, who had a water productivity of 0.42 kg/m^3 only. Our results show that LLT has significant potential to increase water productivity in the Punjab province of India. The higher water productivity basically seems to result from a balanced spread of water on a highly levelled field, which leads to equal supply of water to all plants with no over- or under watering across the patches, as is the case with uneven fields.

7.4.2 Social and Economic Characteristics

Tables 7.4, 7.5, and 7.6 show the socio-economic characteristics of current adopters of LLT users to design practical policies to encourage the use of such systems. The age of farmers did not show any significant variation among farmers under both the LLTs and Non-LLTs users. This finding indicated that the age of a farmer did not

Table 7.4 Social and economic characteristics of respondents using LLT and non-LLT users

Particulars	LLT users		Non-LLT users	
	Number	Per Cent	Number	Per cent
Religion				
Hindu	0	0	0	0
Sikh	36	100	20	100
Other	0	0	0	0
Social category				
General	21	58.33	10	50
SC	6	16.66	4	20
OBC	4	11.11	3	15
EWS	5	13.88	3	15
Other	0	0	0	0
Gender				
Male	36	100	20	100
Female	0	0	0	0
Age (years)				
Less than 30	8	22.22	10	50
30–50	10	27.77	2	10
50–70	4	11.11	2	10
More than 70	14	38.88	6	30

Source: Primary Survey

Table 7.5 Family profile of the respondents using LLT and non-LLT users

Particulars	LLT users		Non-LLT users	
	Number	Per Cent	Number	Per cent
Family type				
Nuclear	32	88.88	17	75
Joint	4	11.11	3	15
Family size				
4 or less	32	88.88	18	90
5–8	2	5.55	1	5
More than 8	2	5.55	1	5
Marital status				
Married	35	97.22	18	90
Unmarried	0	0	0	0
Other	1	2.77	2	10

Source: Primary Survey

affect the choice of irrigation system. But, within the LLT users, it is important to note that about 22% were in the age group of less than 30 years. It means it could also be interpreted that the younger lot of farmers have started realizing the benefits of better techniques and this trend is all likely to go up as the demonstration impacts are increasing with the better spread of social media, etc. The household size of respondents ranged from seven to eight persons for all types of farmers, so the

Table 7.6 Education and occupation profile of the respondents using LLT and non-LLT users

Particulars	LLT users		Non-LLT users	
	Number	Per Cent	Number	Per cent
Education level				
Illiterate	6	16.66	2	10
Up to primary	2	5.55	2	10
Up to matric	8	22.22	10	50
Up to secondary	9	25	2	10
Up to graduate	7	19.44	2	10
Up to postgraduate	4	11.11	2	10
Other	0	0	0	0
Occupation				
Farming	33	91.66	17	85
Business	1	2.77	1	5
Government job	1	2.77	1	5
Private job	1	2.77	1	5
Other	0	0	0	0

Source: Primary Survey

family size was also found to be neutral in determining farmers' attitudes toward the choice of irrigation technology. The education of farmers revealed that farmers using LLTs were relatively more educated than those who used non-LLTs to irrigate their crops. Therefore, education might be a factor affecting the farmers' choice of LLTs. This factor is quite noticeable and would exert more pressure in the coming years as they would better understand the potential of new technology by interacting more with peer groups.

7.5 Conclusions and Policy Recommendations

Punjab may be a water-scarce state, and its agribusiness segment needs water to bolster the quickly growing population. Moreover, the water demand for non-agricultural employment is anticipated to grow more than that of its rural employment. In this manner, a shift to water-saving technologies is required to preserve the balance within the utilization of this precious resource in its competitive employments. Advanced water system frameworks, e.g., LLT, give various other benefits in conjunction with the gigantic water-saving potential. The moo appropriation of these innovations in Punjab could be due to a few restricting variables. Ranchers in Punjab are hesitant to embrace the innovation since they might see it as a costly innovation that gives water investment funds, as it were. In any case, the wasteful water estimating framework fails to motivate the agriculturists to embrace modern water-saving advances for the sole reason of water-saving. The agriculturists might be incentivized through the financial benefits for the appropriation of the LLT. Based on the discoveries of this study, it is concluded that advanced water system frameworks give significant increments in water efficiency of wheat ranches. The outcome of our consideration was that most of the agriculturists utilizing LLT were large ranchers. The display of appropriations and other motivations advertised by the government for the adoption of LLT may not be sufficient to energize the small ranchers to receive LLT, mostly due to a need for awareness of the financial benefits of LLT. In expansion, the moo instruction status of ordinary producers found in this context could also be a limiting factor within the appropriation of LLT. The financial investigation of LLT utilizing Benefit-Cost Proportion affirmed the economic feasibility of modern water system advances within the study area. The use of laser levelling has proven benefits in all respects from saving water to reducing cost, improving yield, and ultimately generating better returns. Importantly, the large-scale use of laser levelling in the end will result in huge savings of irrigation water at the macro level for the state and the country. The large-scale adoption, apart from monetary incentives, much depends upon the strength of extension services in the state, which, as of now are not up to mark because of the decline of the public extension services amidst the ongoing resource crunch and change of priorities of the state in favour of the private services which are supposed to fill the vacuum. The extension system makes a better impact in the situation of demonstrating new technologies by the adoption of better platforms. The better return dimension of new technology needs to be more highlighted for the producers so that, in a situation of stagnating farm incomes, they may feel more attracted towards it. The policymakers could use the findings of this study to educate the farmers about the economic benefits of adopting LLT. The study, overall, argued for the adoption of this technology with apparent benefits. Therefore, the government needs to adopt policies and incentives aimed at water use to ensure the sustainable development of the agricultural sector.

Note: The paper is based on the ongoing doctoral research work of the first author. These are preliminary results based on limited samples and, as such, are not to be quoted. The authors benefited a lot from technical discussions held with experts in the area of water resources, and hence, all those deserve special thanks.

References

Agro-Climatic Zones of Punjab (2020) ENVIS Environment Information System, PSCST. June 4. http://punenvis.nic.in/index1.aspx?lid=5617&mid=1&langid=1&linkid=1257

Aulakh KS (2004) Resource conservation and sustainability under Punjab conditions. Natl Symp Resour Conserv Agric Product 46(4):22–25

Bakshi G (2019) Water conservation in India: how to manage water and what to learn from Israel? The Union Minister of Jal Shakti, September 9. Jagran Josh. jagranjosh.com/current-affairs/water-conservation-in-India-water-management-and-what-India-can-learn-from-Israel-1567845801-1

Bharti V (2019) Water Table Declining Alarming, Government Steps go Down the Drain, September 7. The Tribune, Chandigarh. https://www.tribuneindia.com/news/archive/punjab/water-table-decline-alarming-govt-steps-go-down-the-drain-725015

District-wise Yearly Rate of Fall/Rate of Rise of Water Level - June 1984 to 2017. Ground Water Resources of Punjab, June 23, 2020. http://punenvis.nic.in/index2.aspx?slid=5815&mid=1&langid=1&sublinkid=996

Eyhorn F, Mäder P, Ramakrishnan M (2005) The impact of organic cotton farming on the livelihoods of smallholders, evidence from the Maikaal BioRe Project in Central India. Organic cotton research report, pp 1–75

GOP (2011) Environment Statistics of Punjab, ESO. Government of Punjab, Chandigarh. November 4, Official Website. https://irrigation.punjab.gov.in/Water-Resources.aspx

GOI (2015) Sustainable development goals. Government of India, New Delhi

GOP (2016) Statistical abstract of Punjab, Economic and Statistical Abstract of Punjab. ESO, Government of Punjab, Chandigarh

GOP (2017) State irrigation plan Punjab. Punjab Agriculture University, Ludhiana

Razzaq A, Rehman A, Qureshi AH, Javed I, Saqib R, Iqbal MN (2018) An economic analysis of high efficiency irrigation systems in Punjab, Pakistan. Sarhad J Agric 34(4):818–826

Srivastava SK, Kumar R, Singh RP (2009) Extent of groundwater extraction and irrigation efficiency on farms under different water-market regimes in Central Uttar Pradesh. Agric Econ Res Rev 22(1):87–97

Watto MA, Mugera AW (2015) Econometric estimation of groundwater irrigation efficiency of cotton cultivation farms in Pakistan. J Hydrol Reg Stud 4(1):193–211

Water Resources Department (2018) Ground Water Resources of Punjab State (As On 31st March, 2017). Water Resources & Environment Directorate, Water Resources Department, Punjab, Mohali, and Central Ground Water Board, North Western Region, Chandigarh

Priya is a dedicated economist in her field. Currently, she is pursuing a doctorate from Punjabi University, Patiala. Prior to this, she has 7 months of teaching experience as Assistant Professor at G.H.G. Khalsa College, Sudhar, Ludhiana, in 2018. She has completed her Master of Arts in Economics from S.C.D. Govt. College, Ludhiana, in 2017. She earned her Bachelor of Education degree from G.M.T. College of Education, Ludhiana, in 2015, and she earned her Bachelor of Arts from Khalsa College for Women, Ludhiana, in 2014. Her work has been published in an esteemed Springer journal and an edited book. She has also attended and presented papers at various national and international conferences. She has also participated in a research methodology training programme by ICSSR at Punjab University, Chandigarh, in 2023 to enhance her research skills. She has dedicated herself to economic analysis and sustainable development. Her research innovation has made her an expert in the field of economics, rendering her a skilled economist.

Chapter 8
Assessing Income, Expenditure, and Production Assets of Agricultural Households in India

Manpreet Kaur and Neeraj Sharma

Abstract With 66% of the population residing in rural areas (World Bank 2019), India's economy remains primarily based on agriculture, which continues to be the primary source of income for a significant portion of this population. Agricultural production in India has been steadily increasing and reached nearly 290 million tonnes in 2019; however, India still has an undernourished population of approximately 14% (Grebmer et al., Global hunger index: one decade to zero hunger: linking health and sustainable food systems, 2020). The 2019 household-level data from the NSSO on agricultural households and land and livestock holdings in rural India will serve as the basis for this paper's analysis of the average monthly income and expenses on productive assets and crop production. The NSSO gathered data on receipts and expenses related to crop production, animal farming, and non-farm businesses separately during the agricultural year from July 2018 to June 2019, in two halves.

Keywords Agriculture · Unit level data NSSO-2019 · Income and expenditure

The chapters in this book are presented in the "International Conference on Legal and Economic Synergy for an Equitable Society, organised by the University Institute of Laws at Panjab University Regional Centre, Ludhiana, India on February 1st, 2024".

M. Kaur (✉)
Department of Economics, Guru Nanak College, Killianwali (Sri Muktsar Sahib), Punjab, India
e-mail: aulakhpreet11@gmail.com

N. Sharma
Department of Evening Studies, Multi-Disciplinary Research Centre, Panjab University, Chandigarh, India
e-mail: neeraj2a@rediff.com

A. Virk et al. (eds.), *Law, Economy, and Society: Insights into Legal and Economic Issues in India*, Accounting, Finance, Sustainability, Governance & Fraud: Theory and Application, https://doi.org/10.1007/978-981-95-3361-9_8

8.1 Introduction

Economic growth is an increasing trend in various sectors, moving from its earlier low level to a new, higher one. Internationally, the economic progress of one nation influences the welfare of neighboring countries or those connected through trade (Todaro 2000). Income inequality is increasing in many countries, but large emerging economies like India and China are catching up and may help drive global inequality down. The World Inequality Report of 2022 shows the new estimates of global inequality that the poorest half of the worldwide population barely owns any wealth, possessing just 2% of the total, whereas the richest 10% own 76% of all wealth. The Middle East and North Africa (MENA) are among the most unequal regions in the world, whereas Europe has the lowest levels of inequality. India is notably an extremely impoverished and unequal nation, with the richest 1% of the population controlling about one-fifth of the country's total revenue in 2021 and the poorest having only 13%.

The economic reforms and liberalization adopted by India have primarily benefited the top 1% (WIR 2022). Simon Kuznets' (1963) inequality model discovered an inverted U-shaped relationship between inequality in income and per capita GDP growth using both cross-country and time series data. This concept is a pillar of the modernization school of thought, which emphasizes relative variations in income or wealth as a fundamental determinant of differences in the degree to which money is divided equally within nations. Dependency theory is a key second viewpoint in this field of study (Chan 1989).

India is projected to become the world's most populous country, surpassing China by 2027, according to United Nations population estimates. For Indian agriculture, demographic shift creates a challenge as it grapples with feeding this large population, particularly in light of variability in climate and the depletion of natural resources like air, water, and land. It is also reported that India is predicted to feature nearly 273 million people between 2019 and 2050. By keeping the phenomenon of population growth under consideration, it is clear that the food insecurity will become the most significant threat to human beings soon (Tiwari et al. 2020). In the Global Hunger Index (2018) among 119 qualifying countries, India has ranked 103rd with a score of 31.1. With 66% of the population residing in rural areas (World Bank 2019). Agriculture is the backbone of India's economy and the primary source of income for a significant portion of the country's population.

Despite India's steadily increasing agricultural production, which reached nearly 290 million tonnes in 2019, the country still faces an undernourished population of approximately 14% (Grebmer et al. 2020).

The contribution of agriculture to global GDP has remained relatively stable at around 4% since 2000. In 2021, the agricultural sector employed approximately 866 million people, accounting for nearly 27% of the global workforce. By comparison, in 2000, agriculture employed 1043 million individuals, constituting 40% of the workforce (FAO 2022). Despite 27 million people leaving the agriculture sector between 2000 and 2021 in India and 173 million in China, the combined workforce of approximately 190 million individuals in each country still constitutes two-thirds of agricultural employment in Asia and 44% of global agricultural employment in 2021 (FAO 2022). The non-agriculture sector witnessed higher growth than agriculture, resulting in overall national income increasing by 28.4 times, with an average annual growth rate of 4.83%. However, the number of agriculture workers increased from 9.72 crore in 1950–1951 to 25 crore in 2021–2022 (Chand and Singh 2023).

An equal distribution of income has generated interest worldwide, whether in the context of economic growth, the trade-off between poverty and inequality, or the impact of globalization (Basu 2006; Piketty 2014). Dev (2017) reported that the range of worldwide inequality is considerable, ranging from 0.55 to 0.70, and it has significantly increased in the majority of developed nations as a result of the top 10%'s expanding income share. The issue is also considered to be one of the main elements of the UN Sustainable Development Goals, which are supposed to be accomplished in every country by 2030.

8.2 Review of Literature

Shariff and Azam (2009) examined that agriculture is the main source of income and income disparity in rural India, but between 1993 and 2005, its significance drastically declined. Two factors drive the study of rural India, which are firstly, the fact that 72% of people live in rural areas, suggesting that rural income disparities are a significant contributor to total income inequality. Secondly, only rural areas were included in the 1993 study; the goal of the study is to find out how contributions from various income sources have changed between 1993 and 2005. The results imply that creating employment options outside of agriculture can be used as a tool to lessen economic disparities in rural India.

Dey (2015) by using the neoclassical convergence theory, investigated the inequality and convergence of West Bengal districts. During the study period, the convergence as evaluated by log standard deviation and coefficient of variance, increased. District dispersion grew from 7.54 in 1993–1994 to 8.87 in 2007–2008. The calculated unconditional convergence finding also supports the premise that rich districts expand faster than poor districts. The positive and significant correlation coefficient between gross DDP and various socioeconomic variables supports economic reasoning that higher per capita income districts have higher levels of urbanization, literacy, banking infrastructure, per capita bank advances, number of MSMEs, and employment in MSMEs.

Chakravorty et al. (2016) in their research contributed to the knowledge of inequality and income generation in India's agriculture sector. Using descriptive, decomposition, and modelling approaches, they analyse the National Sample Surveys of Agriculture from 2003 and 2013 and estimate income inequality in the industry at the level of the country and its 17 major states. They demonstrate that (a) income inequality in the agricultural sector of India is very high (Gini Coefficient of about 0.6 during the period), (b) about half of the income inequality is explained by the household-level variation in income from cultivation, which is in turn primarily dependent on variation in landownership, and (c) there are significant state-level differences in the structures/patterns of income generation from agriculture.

Chancel and Piketty (2019) in their paper reviewed the trends of Indian income inequality from 1922 to 2015 by using data from national accounts, household surveys, and data on taxes. According to estimates, the top 1% of earners captured less than 21% of total income in the late 1930s, before dropping to 6% in the early 1980s and rising to 22% in the recent period. Currently, income inequality in India is higher than in the pre-independence period. Before 1922, the estimates show that the top 0.1% of income shared varied between 5% and 7% of national income, but 8% nowadays.

Choudhary and Singh (2019) in their study assessed income inequality, which arises due to regional variations in economic opportunity. While taking primary and secondary data, they used Gini coefficient and the Theil Index to explain the inequality across various regions and districts in Punjab and how inequality arises from variations of economic opportunities across regions. Crops cultivation is the main source of income in almost all quintiles, and livestock, wages, and salaries are the potential sources to bridge the inequality gap. Non-farm business sources would be effective in the central and western zones of Punjab. The Theil index revealed that within-group inequality is the principal contributor to total Inequality across agro-climate zones and districts.

Bathla and Kumar (2019), in their comprehensive study, examined the factors contributing to income inequality among agricultural households across 20 major states. The study utilized the Gini index to assess income disparities among these households, both in total and based on farm size. Employing a regression-based inequality decomposition method, the researchers identified key elements driving inequality. Farming and livestock activities are the major sources of overall income in most states. Despite income disparities persisting across different farm sizes, there has been no significant widening of these gaps over time. Key sources of income include nonfarm earnings, land ownership, and farm assets. Interestingly, the positive correlation between household income growth and land size raises concerns about achieving the government's goal of promoting inclusive development.

Tripathi (2016) in his study reviewed the consumption inequality by using regression-based decomposition analysis. A high level of inequality has been evidenced in India, but it has not been highlighted much due to the lack of credible data on income. The focus of the study is to investigate inequality by considering National Sample Survey (NSS) data between the time period of 2004–2005 and 2011–2012 on consumption expenditure in India. According to the study, the major factors responsible or contributing more to inequality are the size of the household, lack of higher education, less use of technology due to lack of skilled labour, and having a rented dwelling unit in urban areas.

Priscilla et al. (2021) in their study using massive household-level data, attempted to analyze the impact of income sources on inequality amid rural households in the North-east Indian states. The findings show that when income levels rise, farm households are inclined to move away from less lucrative economic activities, such as crop cultivation, and towards higher-value ones, such as non-farm commercial activities. Agriculture contributes the most to regional income disparity, but it also has an inequality-reducing effect. According to theil index estimates, intra-state, intra-district, and intra-landholding size inequality are the region's significant contributions to overall inequality.

8.3 Data Sources and Methodology

The National Sample Survey Office (NSSO) conducted household-level surveys in 2019, specifically the Situation Assessment of Agricultural Households and Land and Livestock Holdings of Households in Rural India. These surveys provided detailed information on land ownership, livestock holdings, and operational details of rural households, including those engaged in agriculture. Each household was visited twice during the survey. In 2013, the survey collected data from 35,200 households in the first visit and 34,907 households in the second visit. This paper aims to analyze the average monthly income, expenditure on productive assets, and crop production based on unit-level data from the NSSO 2019 survey.

Objectives:

1. To analyze the average monthly income (in Rs.) from various sources per agricultural household during the period from July 2018 to June 2019, categorized by different land size classes.
2. To examine the average monthly expenditure incurred on productive assets (in Rs) and crop production.

8.4 Discussion

The National Sample Survey Office (NSSO) collected data on receipts and expenditures related to crop production, farming of animals, and non-farm business during the two halves of the agricultural year from July 2018 to June 2019. Using this information, the average monthly income per agricultural household was calculated for the entire agricultural year (July 2018 to June 2019). The calculation involved summing up income from wages/salary, income from land leasing, net receipts from crop production, income from livestock farming, and non-farm business. Net receipts were determined by subtracting total expenses from total receipts for each income source.

The average monthly income of agricultural households from both farm and non-farm activities is shown in Table 8.1. The relationship between land size classes and related sources of revenue is shown in detail in the table. The size classes go from small land holdings, including those that are less than 0.01 hectares, to larger groups that are more than 10 hectares. The size class of land with land less than 0.01

Table 8.1 Average monthly earnings from various sources for each category of agricultural landholding (measured in hectares) per farming household, during the period from July 2018 to June 2019

Size class of land possessed (ha.)	Income from Wages	Income from leasing out land	Net receipts from Crop production	Net receipts from the farming of animals	Net receipts from non-farm business	Total income
<0.01	6435	254	1435	1087	772	9982
0.01–0.04	4491	189	657	349	703	6388
0.05–1.00	3906	76	2042	356	570	6951
1.01–2.00	3697	74	4313	541	613	9189
2.01–4.00	3548	146	7945	600	758	12,997
4.01–10.00	4273	451	16,914	342	472	22,453
10.00+	3943	581	37,639	7087	1162	50,412
All sizes	4063	134	3058	441	641	8337

hectares earned the most revenue from wages, which was followed by net receipts from farming of animals (1087) and crop production (1435). The largest source of revenue for land owners in the size class of 10.00+ ha was net receipts from crop production (37639), which was followed by net receipts from farmed animals (7087) and income from wages (3943). Notably, the 10.00+ size class stands out with the highest total income, reflecting substantial earnings from multiple sources.

Larger landowners may have the capacity to invest in more advanced farming technologies, hire additional labour, and undertake diverse agricultural and non-agricultural ventures. This reflects the economic concept that resource allocation plays a pivotal role in determining the productivity and profitability of economic activities. Larger land sizes generally result in higher income from crop production and also seem to be associated with higher animal farming income.

The percentage distribution of each agricultural household's monthly average income from July 2018 to June 2019 for all size classes of land owned (in hectares) is presented in graphical Fig. 8.1.

In the agricultural year July 2018–June 2019, the survey gathered information on the costs spent (both on purchases and significant repairs/improvements) on a variety of productive assets utilized for crop production, animal activities, and household non-farm business. However, we will talk about the expenditure on productive assets and agricultural households' crop production in this part. Assets that are frequently used in the production process are regarded as productive assets. Among the assets taken into consideration for a farm business were livestock, various kinds of agricultural tools, and machinery. The average expenditure and receipts incurred from the purchase and sale of productive assets used in farm and non-farm business per agricultural household for each size class of land possessed are presented in Table 8.2.

In households with very small landholdings (<0.01 ha), the average monthly expenditure on productive assets is relatively low, totaling 781 Rs. with the majority spent on livestock and poultry (551 Rs). However, their net expenditure is

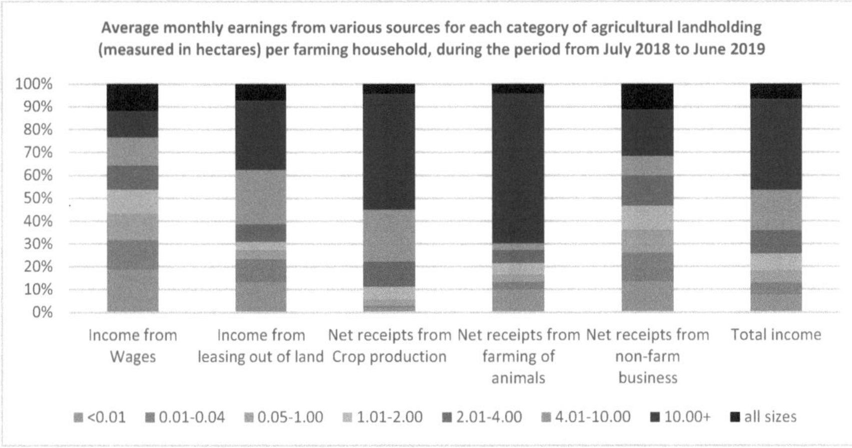

Fig. 8.1 The percentage distribution of each agricultural household's monthly average income from July 2018 to June 2019

significantly high (1691 Rs. for nonfarm business), indicating a possible financial strain. Households with land sizes between 2.01 and 4.00 (ha) spend an average of 1003 Rs. on machinery, which is a substantial increase compared to smaller landholdings, which suggests that larger landholdings might allow for more significant investments in machinery, likely due to higher capital availability and the need for more efficient farming methods.

Households with more than 10.00 + (ha) of land have an average receipt of 7088 Rs. from asset sales, which is significantly higher than smaller landholdings. The net expenditure figures reveal an interesting trend that smaller landholders tend to have higher net expenditures for nonfarm business purposes than for farm business. The figures reflect the increased investment in and reliance on farming as a primary source of income for larger landholders, while smaller landholders might be more dependent on non-farm income sources. The table depicts a correlation between land holding size and financial behavior in agriculture. Larger landholders appear to have more capacity to invest in and profit from productive assets, while smaller landholders face different economic challenges and dependencies (Table 8.3).

The expenses are categorized into different components such as seeds, fertilizers/manures, plant protection chemicals, irrigation, etc. As the size of the land increases, there is a general upward trend in expenses across all categories. Larger land sizes, such as those above 10 hectares, incur significantly higher costs, particularly in areas like lease rent, human labour, and total expenses. The highest expenditure on seeds is observed in the 10.00+ size class (4720), indicating that larger agricultural operations require a substantial investment in planting materials. The 10.00+ size class requires substantial human labour (10709) and employs animal labour (596), indicating the scale and intensity of agricultural activities in larger land holdings. In the context of total expenses, there is a vast difference between 0.01 and 10.00 (ha) and 10.00+ (ha).

Table 8.2 The monthly average expenditure (in rupees) that each agricultural family reports for productive assets and the proceeds from the sale of assets used for both farm and non-farm purposes, reporting the sale of assets that were productive between July 2018 and June 2019

Size class of land possessed (ha.)	Average monthly expenditure incurred on productive assets (Rs)						Receipts from the sale of productive assets (Rs)	Monthly average net expenditure on productive assets per agricultural household
	Used for a farm business				Used for nonfarm business	Total		
	Livestock and poultry	Agricultural machinery and implements	Other productive assets	Total				
<0.01	551	204	26	781	104	855	1691	67
0.01–0.40	121	66	59	246	32	279	8866	−289
0.40–1.00	137	137	223	498	47	545	1702	230
1.01–2.00	186	351	255	792	57	848	2640	401
2.01–4.00	224	1003	798	2026	83	2109	7630	911
4.01–10.00	501	1736	1362	3598	217	3816	2728	2327
10.00+	808	5024	1248	7080	8	7088	6652	4585
All sizes	167	313	273	752	53	806	4559	221

Table 8.3 Average monthly crop-production costs for cultivating agricultural households from July 2018 to June 2019 for each size class of had land (ha)

Size class of land possessed (ha.)	Seeds	Ferilizers/ manures	Plant protection chemicals	Irrigation	Minor repair and maintenance	Interest	Lease rent for land	Labour		All other expenses	Total expenses
								Human	Animal		
<0.01	397	750	267	234	59	77	108	890	25	607	3413
0.01–0.04	134	258	59	101	11	7	35	455	19	212	1290
0.05–1.00	303	520	140	131	22	22	93	896	53	480	2658
1.01–2.00	559	916	316	146	49	59	151	1462	104	881	4644
2.01–4.00	1005	1630	666	180	111	139	525	2459	189	1532	8435
4.01–10.00	1942	2937	1271	253	214	309	1807	4216	295	2993	16,237
10.00+	4720	7515	4299	331	586	1079	14,813	10,709	596	7910	52,559
All sizes	429	715	246	134	40	48	235	1154	74	672	3739

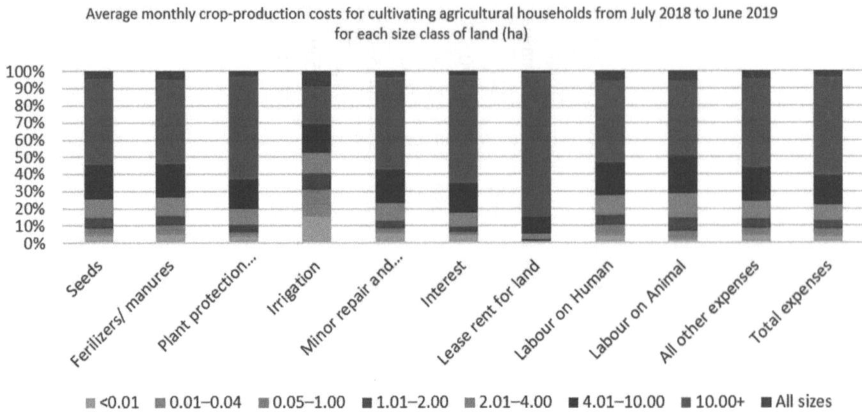

Fig. 8.2 Percentage distribution of monthly expenditure on productive assets by type of asset per agricultural household

A graph presenting the percentage distribution of monthly expenditure on productive assets by type of asset per agricultural household is given in Fig. 8.2.

Implications The above analysis provides a clear insight into the economic dynamics of land ownership and its impact on income sources and expenditures. The larger landowners, with land sizes of 10.00+ acres, have the highest total income, which can be attributed to several factors. Their ability to invest in advanced agricultural technologies and hire labour not only boosts their productivity but also allows them to diversify into different agricultural and possibly non-agricultural ventures. This diversified approach can mitigate risks associated with farming and enhance their income stability. Conversely, the smaller landholders have a different financial strategy or necessity. Their higher net expenditures on non-farm business purposes might indicate a dependency on non-agricultural income sources, which could be due to lower returns from smaller-scale farming. This reliance on non-farm income reflects adaptive strategies to economic constraints, such as investing in small businesses or services that require less initial capital compared to the costs involved in scaling up farm operations.

This trend underscores the economic disparities based on land ownership size, where economies of scale in larger operations provide more opportunities for investment and income generation, whereas smaller landholders face constraints that push them towards other income sources. Such insights are crucial for developing targeted policies aimed at supporting smaller landholders to enhance their agricultural productivity and diversify their income sources, thus improving their overall financial resilience. Through combined efforts such as Cooperative Farming, Adoption of Technology, Diversification, Education and Training, Sustainable Practices, Access to Credit, Renewable Energy Investments, etc., the economic disparities within the farming community can be significantly reduced, leading to more equality.

Conclusion Agriculture employed 866 million people in 2021, nearly 27% of the global workforce, compared with 1043 million (or 40%) in 2000. The 10.00+ size class stands out with the highest total income, reflecting substantial earnings from multiple sources. Larger landowners may have the capacity to invest in more advanced farming technologies, hire additional labour, and undertake diverse agricultural and non-agricultural ventures. The net expenditure figures reveal an interesting trend that smaller landholders tend to have higher net expenditures for nonfarm business purposes than for farm business. The figures reflect the increased investment in and reliance on farming as a primary source of income for larger landholders. In comparison, smaller landholders might be more dependent on non-farm income sources.

References

Basu K (2006) Globalization, poverty, and inequality: what is the relationship? What can be done? World Dev *34*(8):1361–1373

Bathla S, Kumar A (2019) Factors contributing to income inequalities among agricultural households in India. Econ Polit Weekly *54*(21):55

Chakravorty S, Chandrasekhar S, Naraparaju K (2016) Income generation and inequality in India's agricultural sector: the consequences of land fragmentation. Indira Gandhi Institute of Development Research (IGIDR), Mumbai, 4–18

Chan S (1989) Income inequality among LDCs: a comparative analysis of alternative perspectives. Int Stud Q *33*(1):45–65

Chancel L, Piketty T (2019) Indian income inequality, 1922–2015: from British Raj to billionaire raj? Rev Income Wealth 65:S33–S62

Chand R, Singh J (2023) From the green revolution to Amrit Kaal

Choudhary BB, Singh P (2019) How unequal is rural Punjab? Empirical evidence from spatial income distribution. Curr Sci 117(11):1855–1862

Dev SM (2017) The problem of inequality. Rev Dev Change 22(1):1–43

Dey SK (2015) Regional inequality of West Bengal: a district-level study. Bangladesh Dev Stud 38(1):101–117

Food and Agriculture Organization of the United Nations. (2022). The State of Food Security and Nutrition in the World 2022: Repurposing food and agricultural policies to make healthy diets more affordable. FAO. https://doi.org/10.4060/cc0639en.

Kuznets S (1963) Quantitative aspects of the economic growth of nations: VIII. Distribution of income by size. Econ Dev Cult Change 11(2, Part 2):1–80

Piketty T (2014) Capital in the twenty-first century. Harvard University Press

Priscilla L, Singh OK, Vatta K (2021) Effects of income sources on inequality among agricultural households in North-east India. Indian J Agric Econ 76(4)

Shariff A, Azam M (2009) Income inequality in rural India: decomposing the Gini by income sources. Available at SSRN 1433105

Tiwari A, Mishra S, Maurya R (2020) Measurement of inequality and study of social influence on distribution of land holding in India. Int J Agric Sci 16:925–935

Todaro M (2000) Economic development, 7th edn. Addison Wesley, London

Tripathi S (2016) Source of inequality in consumption expenditure in India: a regression-based inequality decomposition analysis

Von Grebmer K, Bernstein J, Alders R, Dar O, Kock R, Rampa F, Wiemers M, Acheampong K, Hanano A, Higgins B, Ní Chéilleachair R, Foley C, Gitter S, Ekstrom K, Fritschel H (2020) Global hunger index: one decade to zero hunger: linking health and sustainable food systems. Concern Worldwide, Bonn: Welthungerhilfe; and Dublin
World Bank (2019) World development indicators. The World Bank
World Inequality Report (2022)

Manpreet Kaur is a dedicated research scholar and a compassionate educator in the field of Economics. She is presently serving as the Head, Department of Economics at Guru Nanak College, Killianwali, District Sri Muktsar Sahib, Punjab (India). Alongside her academic responsibilities, she is pursuing her Doctorate in Economics from Panjab University, Chandigarh (India). With over 14 years of teaching experience, Ms. Kaur has contributed significantly to both undergraduate and postgraduate education, fostering analytical and research-oriented learning among her students. Her academic interests include social security schemes, agricultural economics and income inequality. She has actively participated in numerous national and international conferences, where she has presented more than 30 research papers on diverse socio-economic issues. Her scholarly contributions include several peer-reviewed publications and book chapters in national conference proceedings. As a co-author, she has a book to her credit and has contributed to various academic initiatives focused on applied research and economic policy.

Professor Neeraj Sharma, Head of Economics in the Department of Evening Studies and the Multi-Disciplinary Research Centre at Panjab University, has over 30 years of teaching and research experience at the university. With many firsts to her credit, she is a recipient of the prestigious Shiksha Rattan Award, an award by the Institute of Economic Studies, New Delhi. Her areas of research include rural development and agriculture. She has published more than 50 research papers in reputable national and international journals. She is the author of books titled *Anti-Poverty Programmes in Rural India*, *Dynamics of Rural Non-Farm Sector in India*, and Non-Farm Sector and Livelihood Conditions in Rural Punjab. She has presented numerous research papers and chaired several national and international seminars and conferences. She also organized many workshops and webinars.

Chapter 9
Future Vision and Strategies of Financial Literacy in India with Special Reference to RBI's Initiatives

Kuljeet Kaur

Abstract The need to comprehend money and finances has increased in the modern era due to the fast expansion and complexity of the world's financial systems. It is challenging for ordinary people to make informed financial decisions due to the rapid advancements and constant evolution of technology in the financial sector. For this reason, having financial skills and knowledge—which can be obtained through financial education—is essential. To prevent people from making financial mistakes and maintain financial stability in the country, financial proficiency is crucial. To accomplish the mission of financial inclusion, the Reserve Bank of India (RBI) and other economic regulators in India undertake several financial education campaigns. This paper represents a modest attempt to examine the conceptual underpinnings of the RBI's financial literacy initiatives and to highlight the initiatives led by NCFE for promoting financial literacy, as this needs widespread dissemination. Furthermore, an attempt has been made to recognize the vision and strategies outlined in the National Strategy for Financial Education 2020–25.

Keywords Financial literacy · Financial inclusion · RBI · NCFE · NSFE

The chapters in this book are presented in the "International Conference on Legal and Economic Synergy for an Equitable Society, organised by the University Institute of Laws at Panjab University Regional Centre, Ludhiana, India on February 1st, 2024".

K. Kaur (✉)
PCM SD College For Women Jalandhar, Jalandhar, Punjab, India

Guru Nanak Dev University, Amritsar, India
e-mail: dr.kuljeet.kaur@pcmsd.ac.in

A. Virk et al. (eds.), *Law, Economy, and Society: Insights into Legal and Economic Issues in India*, Accounting, Finance, Sustainability, Governance & Fraud: Theory and Application, https://doi.org/10.1007/978-981-95-3361-9_9

9.1 Introduction

To improve the financial skills of the population, disseminating financial knowledge helps them acquire the necessary skills to make informed financial decisions. Individuals acquire financial knowledge through financial education, which is a key source of how financial literacy is attained.

All societal groups can participate in the government's efforts to financially include all strata of the population in financial programs by possessing strong financial knowledge and skills, which is made possible by financial literacy. The Financial health of any economy is judged by how mainstream institutional entities ensure that low-income and vulnerable groups can obtain the financial services and goods they require fairly and transparently, at an affordable cost. Therefore, financial inclusion makes it possible for the poor to get integrated into the economy's financial system. People must be financially made aware of making wise financial decisions to include vulnerable populations. Financially literate people are well-versed in managing their funds and making correct decisions thereto. This benefits people by boosting both the overall stability of the economy and their financial security.

9.2 Objectives

The main objectives of the paper are

To summarise the initiatives of RBI for enhancing financial literacy and their impact on the financial stability of the economy.
To highlight the Programmes undertaken by NCFE to encourage financial literacy.
To understand the vision and strategies of the National Strategy for Financial Education 2020–25.

9.3 Research Methodology

The research paper is a descriptive type of study. To gain a deep understanding of the subject, a thorough investigation of varied secondary sources, such as research papers, websites, and NSFE reports, was conducted.

9.4 RBI's Initiative on Financial Literacy

RBI has taken various steps to guide individuals to learn about financial decisions. RBI has produced a financial literacy guide, a financial journal, and prepared 16 posters to support people in making the best financial decisions. In addition, since

2016, RBI has held Financial Literacy Week to raise public awareness of making sound financial decisions, savings, well-being, and the appropriate use of online financial services.

The RBI organized Financial Literacy Week with a different subject each year. Financial Literacy Week in 2018 was celebrated with the subject "Consumer Protection," while the theme of Financial Literacy Week 2019 was related to "farmers". Financial Literacy Week 2020 supported the theme "Micro, Small, and Medium Enterprises." The RBI developed a new idea for the Financial Literacy Week topic in 2021: "Credit Discipline and Credit from Formal Institutions." The theme for the Financial Literacy Week in 2022 is "Go Digital, Go Secure." Its objective is to increase public knowledge of safe and secure digital transactions, emphasizing simplicity, lessening consumer anxiety, and safeguarding information.

The RBI has started a massive media effort to impart financial skill lessons under this program. Additionally, banks have been asked to assist in educating their clientele on financial literacy. Additionally, RBI launched the "RBI KEHTA HAI" Special Awareness Campaign to inform the general public. Thus, one of the main roles of the Central Bank is to protect customers through education. The Bank's "RBI Kehta Hai" campaign aims to apprise the public about its initiatives for raising the standard of bank customer care. A knowledgeable consumer can only be a greater guardian of their rights, claims RBI. RBI created the well-known tagline "RBI Kehta Hai, Satark Rahiye, Jaankar Baniye!"

To raise customer awareness, RBI has created a special webpage (rbikehtahai.rbi.org.in) that provides customers with comprehensive information on any pertinent topic, such as Digital Payment Awareness Week, fraud via UPI QR, fraud emails, calls, SMS, and fraudulent transactions. Furthermore, the RBI website's Financial Education portal offers a Financial Awareness Messages (FAME) booklet with a variety of public service announcements. Numerous public awareness campaigns have been executed using alternative media channels, including the RBI's official YouTube channel, a specialized Twitter account called "@RBI says," and a Facebook page called "RBI Says."

Various famous personalities have been made part of the videos related to public awareness campaigns that give the audience an emotional association.

RBI in conjunction with other financial organisations, has also developed the National Centre for Financial Education (NCFE) to advance financial skills for all societal segments in our country in accordance with the Financial Stability and Development Council's national policy for financial education.

9.5 The National Centre for Financial Education (NCFE)

The National Centre for Financial Education (NCFE), is a section 8 (Not for Profit) company under the Companies Act of 2013, which has been promoted by The Central Bank in conjunction with other financial sector authorities such as the Securities and Exchange Board of India (SEBI), Insurance Regulatory and

Development Authority of India (IRDAI), and Pension Fund Regulatory and Development Authority (PFRDA). The main aim of NCFE is to accelerate financial literacy and financial empowerment for all strata of the people by holding workshops, seminars, training sessions, programs, campaigns, and discussion forums across the nation. These events can be held for free or with the assistance of institutions and organizations. Additionally, financial education training is given through the dissemination of study materials, like related literature, pamphlets, booklets, flyers, and technical aids in both electronic and non-electronic formats.

The NCFE develops various programs to encourage financial learning among the masses, including:

9.5.1 The Financial Education Programme for Adults (FEPA)

In 2019, the NCFE introduced a financial literacy program called FEPA which was created and is being used to educate adults, including farmers, women's groups, Asha workers, Anganwadi workers, self-help groups, employees of organizations, and skill development trainees, in many fields of financial management. This program is carried out under the goals of the NSFE, with Special Focused Districts (SFDs) receiving special attention. The NCFE conducts more than 5000 FEPAs each year. It is expected that this program would significantly advance the goal of creating a "Financially aware and empowered India."

9.5.2 Money Smart School Program (MSSP)

This is an effort by the NCFE to offer objective financial knowledge in schools to enhance financial literacy, a crucial life skill for every student's overall development. The program's two main pillars—education and awareness—are meant to build a long-lasting financial awareness that will financially uplift a whole generation. The NCFE encourages educational institutions, especially schools, to incorporate financial literacy curriculum for students in Classes VI through X into their current curricula at their own discretion.

Various Financial Education Workbooks, the study guide for Classes VI to X students, was co-developed by NCFE and CBSE so that it works with the other disciplines taught in the various classrooms. The main advantage for educational institutions that adopt the Money Smart School initiative is that, once their students have gained financial literacy, they will be equipped with the techniques through which they can handle the intricate decisions of their present as well as future financial needs. Moreover, the schools implementing this program will be certified as Money Smart Schools.

9.5.3 The Financial Education Training Program (FETP)

FETP is a tool that NCFE uses to improve financial proficiency in the country by offering unbiased personal financial education to individuals and organizations. FETP is intended especially for educators in India who work with grades 6–10. Participants in this program receive in-depth instruction, and after passing the certification exam, they become "Money Smart Teachers." These trained instructors are then prepared to instruct financial education courses in educational institutions. They also play a part in helping students learn critical financial skills, which helps to further the larger objective of empowering communities through financial literacy. The FETP of NCFE shows a dedication towards financial literacy through the training of educators, who in turn play a significant role in molding the financial awareness of the pupils in their school. NCFE's FETP represents a commitment to establishing a financially informed society.

9.5.4 Financial Awareness and Consumer Training (FACT)

Globally, young people are using credit cards and taking out education loans at earlier ages than in the past. These are big financial decisions that, if not handled carefully, might have long-term effects. Youth must have the skills necessary to set financial objectives, avoid potential pitfalls, and know where to get help when needed as they get ready to graduate and enter the job, taking on greater financial responsibility. It's also critical that they comprehend their obligations and rights as financial customers.

In response to these demands, NCFE launched the Financial Awareness and Consumer Training program, which aims to educate recent young college-going grads on financial matters. This program aims to enhance the financial adoption of this population by covering subjects that are pertinent to them. By providing young people with the information and abilities needed to make wise financial decisions, FACT helps develop a responsible and financially aware generation.

9.6 National Strategy for Financial Education (2020–2025)

To help people manage their money more wisely and achieve financial well-being, NCFE plans to start a massive financial education campaign. To attain the mission objectives, NCFE designs and effectively implements the National Strategy for Financial Education (NSFE). From 2013 to 2018, the first NSFE was implemented. (NSFE: 2013–2018). In collaboration with an outside surveying organization, a survey was carried out in 2019 to highlight the level of financial literacy in India following the conclusion of the NSFE 2013–2018. Comparably, in 2013, as part of the

NSFE 2013–2018 launch, an All-India Financial Inclusion and Financial Literacy Survey was conducted using the OECD-INFE (Organisation for Economic Cooperation and Development–International Network on Financial Education) toolbox. Following a comparison of the findings from these two surveys, NCFE, along with other stakeholders, created NSFE 2020–2025, bearing in mind the rapid advancements in digitalization.

The aim of the new NSFE document is to encourage the progress of the Financial Sector in India. This new document intends to equip various segments of the populace with the knowledge, abilities, attitudes, and behaviours needed to better manage their finances and make future plans. The Strategy advocates a Multi-Stakeholder Approach to attain financial well-being for all Indians.

9.7 Vision of NSFE 2020–2025

The vision of NSFE 2020–2025 encompasses the objectives of its predecessor, NSFE (2013–2018). Its main objectives are:

The aim of financial education is

- To instill financial literacy concepts in all sectors of the population, so elevating it to a valued life skill and promoting proactive saving practices.
- To promote participation in the financial markets to fulfill financial objectives.
- To promote borrowing from official financial institutions when necessary and cultivate credit discipline.
- To educate people regarding safe and secure usage of digital financial services.
- To plan for old age and retirement by investing in suitable pension products and to manage risk at different stages of life with adequate insurance coverage.
- To educate people on their rights, responsibilities, and available channels for resolving complaints.
- To enhance research and assessment techniques used to gauge the growth of financial education.

9.8 Strategies of NSFE 2020–2025

The NSFE 2020–2025 Strategy paper was created with the belief that a one-size-fits-all strategy is inappropriate to achieve the intended goals of financial education promotion. As a result, the 3 D (Dimensions) and 5 Cs approaches form the foundation of NSFE 2020–2025.

9.8.1 The 3 D (Dimensions) Approach

The NSFE's 3 D (Dimensions) method divides the goals of promoting financial education into several areas referred to as Dimensions. The target segment's life stages, with a particular emphasis on the female population, comprise the first dimension. This dimension covers a range of life stages, such as youth, young adults, working adults, and senior persons. This method states that the degree of financial dissemination should differ for all these phases. With an emphasis on vulnerable social groups, geography is the second NSFE 2020–2025 dimension. This dimension advocates providing of varying degrees of financial education to the following populations: hilly states, Lakshadweep, North Eastern Region (NER), aspirational districts, rural and urban areas (with a focus on the urban poor and migrants), and the Andaman and Nicobar Islands. The third dimension is the economic sector, with a concentration on the informal sector, which includes self-employment and the unorganized sector, manufacturing (including skilled and unskilled laborers and craftsmen under the MSME sector and members of SHGs), and agriculture.

9.8.2 The '5 C' Approach

The '5 C' approachalso known as the action plan, is employed under NSFE 2020–25 to disseminate financial education. It places special emphasis on creating pertinent content, building intermediaries' capacity, the community's positive impact, choosing an appropriate communication strategy, and, lastly, improving collaboration among diverse stakeholders. If applied correctly, these Cs have the power to raise the general public's level of financial skills. These are the briefs of all these action plans:

9.8.3 Content

This approach ensures that relevant content is created for a range of stakeholders so that financial education may be delivered correctly. In light of this, it is necessary to create a distinct curriculum or set of instructions for individuals going through various life stages, as described in the first dimension, such as children, young adults, elderly persons, illiterates, and females in particular. Further, it is ensured that the public is provided with relevant content in their preferred language, whether it be regional or simple.

9.8.4 Capacity

This component makes sure that different intermediaries who will aid in the public's financial education are strengthened in their capacity. These intermediaries, who include leaders, bank representatives, master trainers, SEBI resource persons, instructors, counselors, and volunteers, will be informed on the most recent advancements in the financial sector and will instruct the public accordingly.

9.8.5 Community

This action plan promotes the use of local SHGs, teachers, community ambassadors, volunteers, and field-level employees as agents for the spread of financial literacy. The utilization of additional services provided by Anganwadi workers, ASHA workers, Postmen, and others is necessary to encourage people to obtain financial services.

9.8.6 Communication

To spread financial education, this strategy makes use of several cutting-edge communication techniques, such as social media, digital kiosks, bank, websites, and others. Additionally, this entails posting messages in public areas such as bus stops, train stations, hospitals, etc.

9.8.7 Collaboration

To spread financial education, it was determined to work in conjunction with several ongoing projects, as per this action plan. This includes integrating with other programs for teacher training to support the instructors' education, which in turn would support the kids' education. To be given the responsibility of running financial education programs, it was also determined to work in conjunction with Industry Associations, Self-Regulatory Organizations (SROs), and other financial sector intermediaries (such as IBA, FEDAI, AMFI, etc.).

9.9 Conclusion

The poor can engage in the formal financial system thanks to financial inclusion. To incorporate people into the official financial system, financial education about making correct financial decisions is necessary. Financially literate persons are better equipped to handle their money and make educated financial decisions. This boosts the economy overall, along with assisting individuals in becoming more financially stable. Through several initiatives, such as the establishment of NCFE and the observance of financial literacy weeks, RBI plays a major role in educating the public about money and financial decisions. The programs designed to promote financial literacy, including FEPA, MSSP, FETP, and FACT, need widespread dissemination so that the wider population can gain from these programs. The most recent NCFE policy, NSFE 2020–2025, incorporates technology developments, data-driven findings, new policy measures, assessment tools, and interventions. Within the allotted time, the 3D and 5C approaches, each with a distinct goal period, will improve financial literacy. The delivery method and media have been selected on the basis of the needs of the populace since this new approach has learned from experience. The notions of Learning While Doing and Peer-to-Peer Learning, along with efficient mass media outreach, will significantly enhance the financial literacy level of our nation. The goal of NSFE 2020–2025 may be realized through the use of stakeholder collaboration, target audience setting based on life stages, and relatable example utilization. Students in higher education as well as the greater Indian internet community, should be made aware of NCFE's E-learning project. This will enable a wider audience to gain from these resources and raise their level of financial literacy. Furthermore, rather than after a 5-year strategy period, the achievement of NSFE ought to be evaluated after each year.

Bibliography

CFE Teams. https://corporatefinanceinstitute.com/resources/management/financial-literacy. Accessed 15 June 2024

Improving Financial Literacy: Analysis of Issues and Policies I READ online (n.d.) Oecd-Ilibrary. Org. Retrieved from https://read.oecd-ilibrary.org/finance-and-investment/improving-financial-literacy_fmt-v2005-art11-en. Accessed 15 June 2024

RBI's Financial Literacy Week: This Year's Theme Is 'Good Financial Behaviour—Your Saviour'; Know Details (2023, February 13). https://www.outlookindia.com/. https://www.outlookindia.com/business/rbi-s-financial-literacy-week-this-year-s-theme-is-good-financial-behaviour-your-saviour-know-details-news-261863. Accessed 15 June 2024

The History of Financial Literacy (n.d.) OppLoans. Retrieved from https://www.opploans.com/oppu/financial-literacy/history-of-financial-literacy/

Vikaspedia Domains (n.d.). Retrieved from https://vikaspedia.in/social-welfare/financial-inclusion/financial-literacy/national-centre-for-financial-education-ncfe

Dr. Kuljeet Kaur is Assistant Professor in the PG Department of Commerce and Management, PCM SD College for Women, Jalandhar (Punjab). Dr. Kaur has over 14 years of teaching experience and 5 years of dedicated research as Junior and Senior Research Fellow (JRF & SRF). She holds a Ph.D. in Commerce and is a UGC-NET (JRF)-qualified individual, with specializations in Finance and Marketing. She has authored five textbooks for various universities. She has published more than 25 research papers in UGC CARE-listed and peer-reviewed journals, as well as in edited volumes and conference proceedings. Dr. Kaur has actively presented her research at over 20 national and international seminars and conferences. Her academic involvement extends to organizing and coordinating national-level symposia, including an ICSSR-sponsored event for which she also served as editor of the seminar proceedings. With a strong commitment to academic excellence and professional development, Dr. Kaur regularly participates in workshops, conferences, and faculty development programmes (FDPs), making meaningful contributions to the advancement of Commerce and Management education.

Chapter 10
Evaluative Study of Consumer Protection Act with Reference to E-Commerce Transactions

Shikha Puri and Rajni Kapoor

Abstract The proverb "Consumer is King" emphasizes the value of a customer from a marketer's perspective. In order to thrive in the current market, it is imperative that customers' protection be taken into account. Since the emergence of e-commerce, consumer protection has received more attention. Not only has there been a significant shift in the manner that commerce is conducted, but customer awareness has also increased. Mahatma Gandhi remarked, "The most significant guest on our property is the customer. We are depending on him, not the other way around." Since customers generate demand in the commercial market, they are the foundation of the whole supply chain. With only a mouse click, customers may now access a multitude of shopping possibilities while relaxing in the comfort of their own home or workplace, thanks to the widespread use of technology. Customers no longer have to cope with the inconvenience of carrying cash, shopping in crowded areas, or handling salespeople's callous attitudes. Customers are also drawn to websites by the more convenient pricing that is offered there. The current information technology era necessitates appropriate regulatory measures for consumer protection (Goldsmith and McGregor, J Consumer Stud Home Econ 24(2):124–127, 2000). Given the increase in cybercrimes involving online transactions, it is imperative to have an appropriate system in place to address legal and consumer concerns. The notion of consumer protection, its implications, and the shortcomings of consumer protection in the case of e-commerce transactions are the focus areas of this paper. Because of e-commerce trends, special attention has been paid to the laws pertaining to consumer protection that were in effect prior to technological improvements and to different revisions made by the government.

The chapters in this book are presented in the "International Conference on Legal and Economic Synergy for an Equitable Society, organised by the University Institute of Laws at Panjab University Regional Centre, Ludhiana, India on February 1st, 2024".

S. Puri (✉) · R. Kapoor
PCM SD College for Women, Jalandhar, Punjab, India
e-mail: shikhapuri63@gmail.com; rajni.Kapoor1@gmail.com

A. Virk et al. (eds.), *Law, Economy, and Society: Insights into Legal and Economic Issues in India*, Accounting, Finance, Sustainability, Governance & Fraud: Theory and Application, https://doi.org/10.1007/978-981-95-3361-9_10

Keywords Consumer Protection · E-commerce · Laws · Legal provisions · Amendments

10.1 Introduction

Customers are a crucial link in the supply chain. A consumer is defined as any individual or group that makes the ultimate purchase of a good or service offered by many companies. Therefore, consumer protection is critical to the growth of the business sector. E-commerce is becoming more and more popular among customers as more people are connecting to the internet. Electronic commerce, or e-commerce, is the trade of products and services via the internet, involving the transmission of payments and data. E-commerce has surpassed physical commerce in popularity. It is mostly dependent on digital channels such as social media, websites, and mobile applications, which enable buying and selling. The increasing volume of e-commerce transactions indicates the growing acceptance of e-commerce as the primary method of consumer shopping in the current digital era. But the increasing number of users has sparked worries about associated problems like consumer protection (Goldsmith and McGregor 2000; Zwass 2000; OECD 2000). Many issues have been brought up by e-commerce, such as data security, identity verification, customer service, dispute resolution, misleading advertising, return and refund policies, and jurisdictional concerns (Rafat 2007). The protection of consumer rights is challenging as a result of the expansion of electronic commerce. The modern era serves as a canvas on which to depict the fusion of technology and financial sector deregulation (Ramsay 2007). In the current technology era, consumers need to be safeguarded, and laws need to be updated to serve their interests better.

10.2 Research Objective

Consumer Protection laws from antiquity to the present is the focus of this research. It focuses especially on laws pertaining to online consumer security that were passed in 2019 and 2020. The aim of this paper is to examine Consumer Right Protection in the context of e-commerce, draw attention to the challenges in ensuring consumer protection, and offer recommendations for defending consumer interests in this technology age in the Indian context.

10.3 Research Methodology

The Consumer Protection Act of 2019 and the Consumer Protection Rules of 2020 were thoroughly examined for this study (Consumer Protection Act, 2019). The Consumer Protection Act's guiding principles and practical relevance were determined by utilizing a qualitative research approach grounded in text analysis.

10.4 History of Consumer Protection in India

Numerous authors and critics from the past have written about the social, political, and economic aspects of ancient life. Known as the ancient lawgiver, Manu Smriti was the most influential of all of them. He wrote on morality and just trade policies. A code of conduct was established for traders during that era, and those who committed crimes faced less severe penalties. For instance, selling tainted commodities carried a less severe penalty than selling seed grain; if deception was detected, the punishment was mutilation. The king had absolute authority to set trade conditions and determine the appropriate course of action for wrongdoers at this time.

The second significant contribution from ancient India came from Kautilya's Arthashastra, which gave consumer protection a prominent position. It explained the duties of the director of trade, who oversaw the market's state and ensured that ethical business practices were followed.

Consumer protection received the attention it deserved during the medieval era, when leaders such as Alauddin Khilji established stringent regulations and an effective system for enforcing prices. In the contemporary era, the British legal system replaced India's ancient one. Several consumer-related laws were enacted, including the Indian Contract Act of 1872 and the Sale of Goods Act of 1930 (Rafat 2007).

10.5 Indian Consumer Protection Act, 1986

The consumer products industry had experienced tremendous growth and advancement by the 1980s. Thus, the Indian legal system underwent a fresh revolution when the Parliament passed the Consumer Protection Act to safeguard the interests of consumers (Consumer Protection Act 1986). This act's main goal was to deliver justice and easy assessment in a quick, informal, and cost-effective manner. Despite providing a strong legal foundation for addressing consumer complaints, the Act has many shortcomings in light of the current technology environment.

10.6 The 2019 Consumer Protection Act

The Consumer Protection Act of 2019 was ratified by the Indian President on August 9th, 2019 (Consumer Protection Act, 2019). The Act of 1986 was superseded by this one. Its goal was to address the ways in which unethical corporate practices and unfair trade practices could expose consumers to exploitation.

10.7 Consumer Protection Regulations 2020 (E-Commerce)

In order to safeguard consumer rights, the Consumer Protection (e-commerce) rules 2020 were developed (Consumer Protection Rules, 2020). They include information about e-commerce handlers and take into account both the customer and product service providers.

10.8 E-commerce and Consumer Protection

Information and communication technology has completely changed how businesses operate. With new technology developments, e-commerce transactions exhibit innovation. Concerns regarding data privacy, consumer protection legislation, privacy violations for businesses, and government have arisen as a result of the convergence of e-commerce transactions (E-Commerce Group, 2009).

10.9 Threats to Consumer Protection Posed by E-Commerce

Concerns about infringement of consumer rights have been raised by e-commerce. These include predatory pricing, unfair practices, deceptive marketing, internet fraud, and violations of data privacy (Ashar 2000).

10.10 Recommendations for Improvement in Provisions Pertaining to Consumer Protection

Proper regulatory mechanisms and policy definitions must be developed to address unfair trade practices, drop pricing, privacy violations, and data protection. Guidelines must be provided for small and local vendors to participate in the e-commerce ecosystem (Goldsmith and McGregor 2000).

10.11 Limitations of the Study

This research is not an exception to the rule that most studies have some limitations. The primary obstacle is the lack of sufficient literature outlining the effects of the legislative framework governing consumer protection in e-commerce.

10.12 Conclusion

The Consumer Protection Act still has inadequate provisions, and it is too early to comment on how effective these laws are at ensuring consumer safety and security. Over time, legal and regulatory developments, supported by judicial interpretation, can provide a solid foundation for consumer protection in the digital age (Ramsay 2007).

References

Ashar A (2000) Consumer Protection in E-Commerce Report on OECD Guidelines for Consumer Protection in the Context of Electronic Commerce. Published in Asia-Pacific economic cooperation electronic commerce steering group workshop on consumer protection. http://www.accc.gov.au/content/item.phtml

Consumer Protection Act (1986) Consumer affairs, food and public distribution sector. Government of India. http://india.gov.in/sectors/consumer_affairs/index.php

Goldsmith E, McGregor SLT (2000) E-commerce: consumer protection issues and implications for research and education. J Consumer Stud Home Econ 24(2):124–127

OECD (2000) Guidelines for consumer protection in the context of E-Commerce. Available at http://ww.accc.gov.au/content/item.phtml?itemId=255556&nodeId=ee905a658e5458e77a3c4182605e98d2&fn=Asia.pdf

Rafat J (2007) Consumerism and legal protection of consumers. Deep and Deep Publications

Ramsay I (2007) Consumer law and policy: text and materials on regulating consumer markets. Hart Publishing

Zwass (2000) Structure and macro-level impacts of electronic commerce: from technological infrastructure to electronic marketplaces. http://www.mhhe.com/business/mis/zwass/ecpaper.html

Mrs Shikha Puri completed her B.Com and M.Com from Panjab University, Chandigarh. She cleared her UGC-NET in the first attempt and has also completed a Diploma in Statistics. Her specialisation includes accounting and corporate law. With over 16 years of teaching experience as an Assistant Professor in Commerce, she has authored numerous research papers at the national and international levels. She has to her name One Patent Publication. She has also written five books, covering subjects such as accounting, Production and operations management, and risk management. Her hard work, dedication, and zeal for learning have greatly aided her in her journey to becoming an exemplary academic and researcher.

Mrs. Rajni Kapoor is an accomplished academic with over 32 years of experience, currently serving as an Associate Professor in the PG Department of Commerce and Management. She holds an M.Com degree and specializes in Finance. Her professional contributions include authoring a book and publishing 10 research papers, with nine at the national level and one internationally. She has presented papers at six conferences, attended five seminars, and participated in 31 webinars, highlighting her active engagement in academic discourse. Mrs. Kapoor has also attended two national workshops, four orientation/short-term courses, and 12 faculty development programs. She has successfully completed a UGC-sponsored minor research project on consumer rights. She has been actively involved in academic bodies, including the Board of Studies and the Faculty of Economics and Business. Her accolades reflect a dedicated career in education, research, and institutional service.

Part IV
Economic and Legal Dimensions of Sustainability

Chapter 11
Environment, Law, and Livelihood of Marginalized Occupations: A Case Study of Pottery Artisans

Dikshika Jakhar, Jaspal Singh, and Anjali Bansal

Abstract Pottery has deep roots in Indian culture and tradition, dating back thousands of years. Many regions in India have their distinct pottery styles and techniques, reflecting the country's diverse cultural heritage. Pottery plays a vital role in preserving cultural heritage, supporting livelihoods, promoting economic development, and fostering artistic creativity. Artisans employed age-old techniques and skills to create a wide variety of pottery items, including earthenware, terracotta, ceramic, and porcelain products. It serves as a livelihood for many people, particularly in rural areas. The industry generates employment opportunities for skilled craftsmen, providing economic sustenance to numerous families and communities. The pottery industry faces a critical challenge of striking a balance between environmental preservation and sustainable practices. The research paper examines the complex relationship between environmental protection regulations and the sustainability of pottery production processes. This study examines the current regulatory framework governing the ecological aspects of pottery making. The major problems, including emissions control, waste management, and resource conservation,

The chapters in this book are presented in the "International Conference on Legal and Economic Synergy for an Equitable Society, organised by the University Institute of Laws at Panjab University Regional Centre, Ludhiana, India on February 1st, 2024".

D. Jakhar (✉)
SGT University, Gurugram, India
e-mail: dikshika.jakhar@gmail.com

J. Singh
Consultant NITI Aayog, New Delhi, India
e-mail: punjabimatti82@gmail.com

A. Bansal
Consultant NITI Aayog, New Delhi, India

IIT Roorkee, Roorkee, India
e-mail: anjali.bansal292@gmail.com

A. Virk et al. (eds.), *Law, Economy, and Society: Insights into Legal and Economic Issues in India*, Accounting, Finance, Sustainability, Governance & Fraud: Theory and Application, https://doi.org/10.1007/978-981-95-3361-9_11

are reviewed within the context of environmental regulations and their implications for the pottery sector. Moreover, the paper assesses the strategies that the government should adopt to comply with environmental standards while maintaining economic viability and competitiveness. This research contributes to the discourse on achieving a harmonious balance between environmental protection and the sustainable development of the pottery-making industry.

Keywords Pottery industry · Manufacturing · Artisans · Environment · Law · Livelihood

11.1 Introduction

The nexus between environmental considerations, legal frameworks, and the livelihoods of marginalized communities presents a complex matrix of challenges and opportunities that are crucial for achieving Sustainable Development Goals (SDGs). This chapter delves into the intricate world of pottery artisans, a marginalized occupational group, whose traditional practices are deeply intertwined with ecological processes and existing legal mechanisms. By examining the case of pottery artisans, we explore how the interplay between environment, law, and livelihood not only affects their economic stability and cultural continuity but also impacts broader sustainable development objectives. India is a country immersed in a diverse array of cultures, where ancient crafts and artistic traditions have flourished for many centuries. Artisans in India represent a wide spectrum of crafts and expertise and have served as guardians of the country's legacy, and play a pivotal role in shaping its identity. Artisans in India serve as a significant pillar of culture and heritage. These artisans contributed to the country's social, economic, and cultural tapestry. Moreover, the development of rural is concerned with growth of small-scale industry. Among various small-scale industries, pottery industry is of great importance in growth and development of rural economy (Pegu 2019). Pottery, one of the oldest human crafts, embodies a unique blend of cultural heritage and practical utility. Artisans engaged in pottery often rely on natural resources, such as clay, water, and firewood, which links their livelihood directly to the environment. The sustainability of these resources is pivotal, not only for the continuation of their craft but also for the ecological health of their communities. Therefore, understanding the environmental impact of pottery is essential in assessing the sustainability of these practices under the current global emphasis on ecological conservation and sustainable use of natural resources (United Nations Environment Programme 2019).

The evidence of pottery artisans can be traced from the period of Harappan civilization (Rao and Lal 2010). The Indian ceramic industry has a striking picture (Kiradoo 2021). There is rising demand of ceramic products in a domestic as well as international market. However, the livelihood of these artisans encounters various obstacles as Indian ceramic industry is ready to tackle global competition undergoing the phase of modernization through cutting edge innovation in products, quality, and design (Messe 2018). However, the traditional methods being used industry specially the use of conventional kilns, have come under the increasing scrutiny due to their contribution to air pollution. The production processes involved in ceramic manufacturing from raw material to finished product, pose the multiple environmental risk to the planet (Ganesh 2019). Gas emissions originating from the ceramic industry pose significant environment risks. These emissions stem from the storage and handling of raw materials as well as during drying processes with hot air or firing. They can manifest as particles, atoms and gasses primarily generated during firing process (Ibrahim 2016).

Moreover, legal frameworks governing resource access, environmental protection, and workers' rights play a significant role in shaping the livelihoods of pottery artisans. In many regions, artisans face regulatory challenges that may either support or hinder their craft. For instance, laws regulating the extraction of clay can significantly affect the availability of raw materials, thus impacting production and livelihood (Smith 2015). Additionally, environmental laws designed to protect ecosystems can inadvertently limit the traditional practices of artisans unless specific provisions are made to accommodate these livelihoods within the legal structures.

As the concern for environment conservation escalate, regulatory bodies, such as the National Green Tribunal (NGT), have stepped in to address the challenges being by air pollution. The NGT operates as a specialized court with the broad authority, comprising judges and environmental experts. Section 20 of the National Green Tribunal Act 2010 requires adherences to key principles of international law, including sustainable development, precautionary measures, and the "polluters pay" principle (Gill 2014). These artisans are contending various regulatory and market forces, potentially imposing significant financial and technological strains on them (Panghal et al. 2023). So, the pottery industry, being deeply rooted in tradition and artisanal techniques, finds itself at a cross blade as it grapples with the complexities of regulatory compliance. Compliance with NGT regulations requires huge investment in technology upgrades and pollution control equipment. However, the per day earning of workers engaged in traditional sector is low as compared to the other sectors of economy (Jakhar et al. 2023). The financial burden associated with these requirements often exceeds the means of small-scale artisans threatening their livelihoods and traditional practices.

The livelihood of pottery artisans is not just a function of their ability to access and utilize natural resources but also of the economic viability of their craft in contemporary markets. Globalization and the influx of mass-produced goods have posed significant threats to traditional crafts (Doe and Patel 2017). Thus, understanding the market dynamics and the economic pressures faced by artisans is crucial for developing strategies that enhance their livelihoods without compromising environmental and cultural sustainability.

Integrating these factors, this chapter argues that a holistic approach encompassing environmental sustainability, supportive legal frameworks, and economic empowerment is essential for improving the lives of pottery artisans and contributing to the achievement of SDGs. Specifically, SDGs such as Goal 1 (No Poverty), Goal 8 (Decent Work and Economic Growth), Goal 12 (Responsible Consumption and Production), and Goal 15 (Life on Land) are directly relevant to the pottery sector (United Nations XE "United Nations" 2021). By ensuring sustainable practices, fair legal support, and market opportunities, not only can the livelihoods of pottery artisans be secured, but a significant stride can also be made towards achieving these global goals.

Furthermore, the intersection of environmental sustainability, legal frameworks, and livelihoods provides a critical lens through which the challenges and opportunities faced by pottery artisans can be examined. For instance, enhancing legal recognition of artisanal rights, coupled with training on sustainable practices, could lead to improved livelihoods while contributing to environmental goals (Johnson and Anderson 2018). Additionally, policies that integrate market support mechanisms with environmental and social sustainability could create more robust systems for preserving traditional crafts and ensuring the economic well-being of artisans (Brown and Green 2020).

This chapter employs a multidisciplinary approach, drawing from environmental science, law, and socio-economic studies, to provide a comprehensive analysis of how intertwined factors affect the livelihoods of pottery artisans. Through this analysis, we aim to offer insights into the broader implications of these intersections for sustainable development and the potential pathways through which marginalized communities can enhance their living standards while contributing to the global sustainability agenda.

11.2 Objectives

This research paper investigates the legal framework surrounding air pollution from the use of conventional kilns in pottery making. Further, it focuses on the ongoing discourse on environmental law and sustainable development by shedding light on pottery making. The paper also discusses various challenges faced by the pottery industry and suggests policy recommendations that align with the principles of environmental justice and sustainable development. This study serves as a bridge between the regulatory measures implemented by the authority and the lived experience of the pottery artisans.

11.3 Environment Issues and Pottery Makers

The pottery industry, celebrated for its exquisite craftsmanship and timeless creations, casts a shadow on the environment that cannot be overlooked. Behind the allure of delicate pottery and gleaming tiles lies a harsh reality of environmental degradation and ecological imbalances. Extraction of raw material can lead to habitat destruction, soil erosion, and disruption in the ecosystem. Further, firing pottery contributes to carbon emissions leading to air pollution. Air pollution emerges as a critical concern associated with ceramic industry and these air pollutants causes severe environmental issues and becomes a significant health risk for livelihood of artisans (Tripathi et al. 2019). Production, pollution, and protection are inter-related terms. The ceramic industry is similarly intertwined with this connection. The traditional kilns used for bonfire kilns, characterized by open firing in shallow pits. Smoke generated during firing poses significant risks to environment (Govind et al. 2021).

The emissions degrade air quality and possess health risks to nearby communities, particularly those living in the proximity to pottery making facilities. As we delve deeper into the intricacies of this industry, a narrative of unsuitable practices and consequently impacts unfolds, urging to reevaluate sustainable alternatives.

11.4 Law to Protect of Environment Issues

Addressing the environmental challenges posed by ceramic industry requires a multifaceted approach that encompasses technological innovation, policy reforms and industry-wide collaborations. In India, environment regulations are primarily governed by the Ministry of Environment, Forest, and Climate Change (MoEFCC). The Air (Prevention and Control of Pollution) Act, 1981 and the Environment (Protection) Act, 1986 are some of the key legislations related to air pollution control. There are different legislations being imposed by this regulatory authority to curb the air pollution. Industries are required to obtain the consent to operate from the State Pollution Control Board (SPCB) or Pollution control Committee (PCC) before starting these operations. This involves compliance with environment norms, including those related to air pollution. The Central Pollution Control Board (CPCB) sets the emission standards for various industries to control the release of pollutants into the air. Industries, including pottery units, are expected to adhere to these standards. Moreover, proper disposal of waste industrial waste including pottery waste is a crucial aspect of environment regulations. Industries are expected to manage and dispose of their waste in an environmentally friendly manner.

Further, larger pottery units of those located in ecologically sensitive areas may be subject to environmental impact assessments to evaluate potential environmental consequences. These stringent requirements often disproportionately impact small and medium-scale enterprises (SMEs) like pottery industry. As a result, traditional potters are abandoning their craft due to insufficient income to sustain themselves and compliance with these excessive regulations. Further, the National Green Tribunal in Delhi recently mandated the closure of all pottery kilns in Uttam Nagar (Dahiya et al. 2021). Switching to the new technology requires greater financial resources which is difficult for small firms to afford.

11.5 Livelihood of Pottery Artisans

Artisans play a vital role in the economic advancement of the nation. Artisans held esteemed positions and were esteemed for their artistic contributions within the society (Verma and Gupta 2017). Maximum number of artisans are still living a traditional way of living (Akilandeeswari and Pitchai 2016). Pottery industry is significant it affects the economic and cultural perspective of the nation. It helps in generation of employment, raising exports and reducing regional disparities (Regon 2019). Pottery making relies on specific raw material, tools, expertise, and knowledge (Pfaffenberger 1992). The ceramic industry in India produces different categories of products which includes sanitary ware, refractories, cement, advance ceramics, and ceramic tiles (Biswas 2019). The ceramic industry is an important industry which affects are performance of the other primary industries (Mandal 1984). Moreover, Ceramic industry provides an employment to the large number of populations. However, artisans in the ceramic industry have encountered the various challenges which affects their socio-economic livelihoods (Kasemi 2014). This section highlights the trend of ceramic workers in India and their daily income over the years.

Small and Medium enterprises lead to socio-economic development of nation, especially in terms of employment and job creation (Amoah and Amoah 2018). Further, Ceramic industry being an upcoming sector due to increased demand domestically and globally, sustains the livelihoods of approximately 4.5 lakhs individuals by offering employment opportunities. Table 11.1 shows the changing trends of working population and the ceramic workers in India from 1993 to 2021. There was a declining trend in workforce participation rate in India after 1993–1994 in the data of employment and unemployment survey. However, after replacing the E&U with periodic labor force survey in 2017–2018, the turned out to be reversible. as shown in Table 11.1. There was a continuous increase in the population except in 2017–2018. The picture of ceramic sector was just the opposite.

Table 11.1 Changing trends in ceramic workers in India

Year	Population (in millions)	Worker population ratio (%)	Number of worker (in million)	% of Workforce in ceramic product manufacturing	Ceramic worker (in lakh)
1993–1994	895	42.0	376.0	0.2353	8.85
2004–2005	1093	42.0	459.1	0.1725	7.92
2011–2012	1227	38.6	473.8	0.1218	5.77
2017–2018	1314	34.7	456.0	0.1007	4.59
2018–2019	1327	35.3	468.4	0.1015	4.75
2019–2020	1341	38.2	512.3	0.1001	5.13
2020–2021	1355	39.8	539.7	0.0920	4.52

Source: Jakhar, D., Sidana, N., Singh, J., & Singh, S. (2023). Ceramic Manufacturing Workforce in India: Trends, Relative Earnings and Its Determinants. *Transactions of the Indian Ceramic Society*, 82(4), 271–277

Table 11.1 showed that there was continuous decline in the share of ceramic workers in total workforce in India. In 1993–1994, the share of ceramic workers in total workforce was 0.23% which was very low. In 2004–2005, although the working population showed the rising trend, but ceramic workforce got declined. Further, in 2018–19 still there was rise in workforce participation population but ceramic workforce shows the declining trend. Similar pattern was being traced over the years. The declining workforce in ceramic sector amidst as an industry of potential underscores the complex interplay of technological, economic, and social factors shaping workforce dynamics.

Table 11.2 displayed the annual growth rate of both total workforce and ceramic workforce from 1993 to 2021 in India. There was a continuous increase in the total workforce from 1993 to 2019 except in 2011–2018, however in 2019 there was reduction from 9.36% to 5.36% due the covid pandemic breakout. On the contrary, the growth rate of ceramic workforce declined from −2.7% to −3.12% during 1993–2018. It further declined to −8.07% in 2020–2021. Also, the ceramic workers faced significant declined during the covid pandemic. In 2019, ceramic industry has experienced a hazy period because of the obstacles in its user (real estate) sector prior to onset of Covid-19. Although, ceramic industry is an emerging industry, one would expect a corresponding rise in the number of individuals employed in this sector, but there was a drastic decline in the workforce engaged in this industry over the period. This shows that people are reluctant to work in this sector, which reflects a complex interplay of different variables like economic, technological, regulatory, and cultural factors.

Income being the important factor that affects the upliftment of the different segment of the society (Purcell et al. 2016). The progress of nation depends on the level of income of the individuals. Table 11.3 highlighted the average per day earning of ceramic and non-ceramic workers over the time in India. There was a wide gap in per day earnings of ceramic and non-ceramic workers. In 1993–94, the per day earning of other worker and ceramic worker was Rs. 22 and Rs. 21 respectively. There was not much gap in the earnings during 1993–1994. In 2004–2005, the gap between the ceramic and other workers per day earnings has widened up to Rs. 18. In 2011–2012, the per day earning gap rose to Rs. 21.

Table 11.2 Annual growth rate of ceramic and overall workforce change over time

	Growth rate (%) per annum in workforce	
Time duration	Total workers in economy	Workers in ceramic sector
1993–1994 to 2004–2005	1.83	−2.78
2004–2005 to 2011–2012	1.06	−3.12
2011–2012 to 2017–2018	−0.64	−3.12
2017–2018 to 2018–2019	2.74	0.81
2018–2019 to 2019–2020	9.36	−1.37
2019–2020 to 2020–2021	5.36	−8.07

Source: Jakhar, D., Sidana, N., Singh, J., & Singh, S. (2023). Ceramic Manufacturing Workforce in India: Trends, Relative Earnings and Its Determinants. *Transactions of the Indian Ceramic Society*, 82(4), 271–277

Table 11.3 Average earnings of ceramic workers versus non-ceramic workers in India over time

	Per day earning (Rs.) at current prices		
Year	Non ceramic worker	Ceramic worker	Ratio (Non Ceramic/Ceramic)
1993–1994	21.9	21.0	1.04
2004–2005	93.4	75.1	1.24
2011–2012	191.9	171.1	1.12
2017–2018	311.4	235.3	1.32
2018–2019	339.5	271.6	1.25
2019–2020	329.4	246.6	1.34
2020–2021	340.2	256.7	1.33

Source: Jakhar, D., Sidana, N., Singh, J., & Singh, S. (2023). Ceramic Manufacturing Workforce in India: Trends, Relative Earnings and Its Determinants. *Transactions of the Indian Ceramic Society*, 82(4), 271–277

From 2004 to 2021, there was a continuous increase in the gap between the daily earnings of ceramic and non-ceramic workers. In 2020–2021, the difference between the per-day earnings of ceramic workers and other workers was Rs. 83. This highlighted the inequality that persisted within various sectors. The widening gap between the daily earnings of ceramic and other workers was discouraging for the ceramic workers from entering this sector. Despite their skilled labor and dedication, they find themselves trapped in low-paying jobs with limited prospects. This undermines the economic stability of their family.

11.6 Contradiction Between Environmental Protection Laws and the Survivable of Marginalized Communities

On one hand, there is a promotion of "Make in India", and on the other, we are on the brink of losing their source of income from the marginalized industry, like the pottery industry. Balancing economic growth initiatives like "Make in India" with preserving traditional industries like the pottery industry is challenging. The pottery industry, often rooted in local craftsmanship and culture, represents an important part of India's heritage and provides livelihood for many marginalized communities. Regulations are necessary for safety and quality standards, but they must also be designed in a way that supports small-scale industry rather than stifling it. Environmentally friendly alternatives for wood-fired kilns are either an electric oven or an oven run on LPG. The installation and maintenance cost of new equipment is a cause of worry for small pottery families, which will increase the production cost of pots and other clay and ceramic products in the market. Already this industry is facing stiff competition from the cheap products in market like plastic (Lakshman 1966).

Moreover, there is a glaring gap between the daily earnings of the ceramic and non-ceramic workers. Ceramic workers do not yield sufficient returns to sustain a decent livelihood. Excessive regulations not only impact their own well-being but also undermine the economic stability of their families and communities, making these artisans reluctant to enter this profession. Regulatory requirements pose hurdles for these artisans. From obtaining a license and permits to complying with health and safety standards. The burden of regulatory compliance adds another layer of complexity to an already challenging landscape and dissuades artisans from pursuing their passion for pottery making, pushing them to the brink of extinction, which creates a tension between economic progress and cultural preservation. Pottery artisans are ready to cooperate if the government can provide monetary help in the setup of new, environmentally friendly kilns, provide alternative fuel options on a subsidy, or provide alternative jobs to their families.

Therefore, a balance between environmental protection and preservation of traditional craftsmanship, especially for small pottery artisans, is needed. It is crucial for the government to consider the socio-economic impact of regulations and ensure that they don't harm these vulnerable communities, and focus on raising awareness about the importance of these industries and their sustainability.

11.7 Conclusion and Suggestive Way Forward

Pottery, an ancient craft deeply ingrained in cultures around the world, holds a special place in the narrative of human history. Traditional practices often involve resource-intensive processes, such as wood-fired kilns emitting pollutants and consuming a vast amount of fuel. In today's era of heightened environmental awareness, addressing these ecological concerns is paramount. Yet, the transition towards environmentally friendly practices must be approached with sensitivity to the socio-economic realities of small pottery artisans. Abrupt regulatory changes or the enforcement of stringent environmental standards without considering the capacity of these artisans could jeopardize their livelihoods, leading to economic hardship and the loss of invaluable cultural heritage.

Government plays a crucial role in mediating this delicate balance. Engaging with policymakers, raising awareness about the importance of these industries, and advocating for policies that support their sustainability. Establishment of a phased approach for small artisans to meet emission standards, allowing them sufficient time to adjust without immediate financial strain. The government should focus on allocating funds for research and development initiatives to discover and promote low-impact materials and production processes in pottery making. Establishment of community-based waste management systems to support small artisans in adhering to environmental standards. Moreover, there should be the implementation of separate regulations for small and large enterprise units. Provision of financial incentives and support for small artisans to adopt cleaner and more efficient technology. Further, the provision of training programs will help the artisans in transition to eco-friendly practices without compromising their traditional craftsmanship.

In conclusion, addressing the challenges faced by pottery artisans requires a nuanced understanding of the environmental, legal, and economic contexts in which they operate. By fostering an integrated approach that considers these various dimensions, it is possible to design interventions that not only enhance the livelihoods of these artisans but also promote broader sustainable development outcomes. This commitment is echoed in the Government of India's dedication to preserving traditional livelihoods, exemplified by initiatives such as the Vishwakarma Scheme, which aims to support artisans by providing training, financial assistance, and market access, thereby ensuring that the cultural and economic fabric of traditional crafts is maintained (Government of IndiaXE "Indias" 2022). This case study not only highlights the specific needs and challenges of pottery artisans but also serves as a model for understanding and supporting other marginalized occupational groups in a way that aligns with the objectives of the SDGs.

So, balancing environmental protection with preservation of traditional craftsmanship for small pottery artisans is crucial for safeguarding the cultural heritage and supporting economic livelihood, and promoting environmental sustainability.

References

Akilandeeswari SV, Pitchai C (2016) Pottery industry and its development by effective marketing through information and communication technologies (ICT). Int J Res 4(4):6–13

Amoah SK, Amoah AK (2018) The role of small and medium enterprises (SMEs) to employment in Ghana. Int J Bus Econ Res 7(5):151–157

Biswas B (2019) The relation between capital structure and profitability of some selected companies in the Indian ceramic industry. IUP J Account Res Audit Pract 18(3):49–58

Brown A, Green T (2020) Market integration and sustainability in traditional craft sectors. Oxford University Press

Dahiya S, Siwach P, Panghal A, Sindhu S (2021) 11 terracotta pottery catastrophe. Entrepreneurship and Big Data: The Digital Revolution, 153

Doe J, Patel S (2017) Globalization and traditional industries: impacts and responses. Cambridge University Press

Ganesh C (2019) The environmental impact caused by the ceramic industries and assessment methodologies. Int J Qual Res 13(2):315–334

Gill GN (2014) The national green tribunal of India: a sustainable future through the principles of international environmental law. Environ Law Rev 16(3):183–202

Government of India (2022) Vishwakarma scheme: empowering artisans. Ministry of Skill Development and Entrepreneurship

Govind K, Keerthidas PU, Karthik R, Kishore R (2021) Design and simulation of pottery kiln. Research Square. https://doi.org/10.21203/rs.3.rs-593830/v1

Ibrahim NA (2016) Risks caused by ceramic industry emissions and methods of treatment. Int Design J 6(2):263–270

Jakhar D, Sidana N, Singh J, Singh S (2023) Ceramic manufacturing workforce in India: trends, relative earnings and its determinants. Trans Indian Ceramic Soc 82(4):271–277

Johnson L, Anderson M (2018) Legal frameworks and sustainable development: a study of artisan rights. Law Soc Rev

Kasemi N (2014) Problems of pottery industry and policies for development: case study of Koch Bihar district in West Bengal, India. Int J Adv Res Manage Soc Sci 3(7):238–247

Kiradoo G (2021) The transition of traditional pottery-making into advanced ceramics in context to the Indian ceramic industry. Turk J Physiother Rehabil 32(2):546–556
Lakshman TK (1966) Cottage and small scale industry in Mysore. Rao and Raghavan. Mysore, pp 129–149
Mandal G (1984) Ceramic education in India and its growth potential. Trans Indian Ceramic Soc 43(5):140–144
Messe Muenchen India and EAC International Consulting (2018) Status quo and outlook 2022: Indian ceramics industry. Ceramics Asia, Mumbai
Panghal A, Dahiya S, Sindhu S, Siwach P, Dahiya B (2023) Strategic business model canvassing for terracotta pottery entrepreneurs in India. Int J Entrep Small Bus 50(3):407–432
Pegu D (2019) Pottery industry in Majuli District of Assam: it's problems and prospect. Int J Sci Technol Res 8(12):3650–3652
Pfaffenberger B (1992) Social anthropology of technology. Annu Rev Anthropol 21(1):491–516
Purcell SW, Ngaluafe P, Foale SJ, Cocks N, Cullis BR, Lalavanua W (2016) Multiple factors affect socio-economic and well-being of artisanal sea cucumber fishers. PLoS One 11(12):1–20
Rao TY, Lal BS (2010) Rural artisans-indigenous technology: an empirical study on village potters in Warangal. Indian J Dev Res Soc Action 5(1):309–317
Regon BJ (2019) Problems and prospects of pottery industry in Majuli. Int J Res Eng Sci Manage 2(5):952–955
Smith R (2015) Resource management and legal systems: a case study of clay artisans. J Environ Law Policy
Tripathi CB, Baredar P, Tripathi L (2019) Air pollution in Delhi. Curr Sci 117(7):1153–1160
UNEP (2019) Sustainable resource management and traditional crafts. United Nations Environment Programme
United Nations (2021) The sustainable development goals report 2021. United Nations
Verma A, Gupta (2017) I. A study on working condition of artisans in Saharanpur wood craft industry. J Art Architect Stud (JAAS) 6(1):12–17

Dr. Dikshika Jakhar obtained her B.Com. (Programme) degree from Daulat Ram College, University of Delhi, in 2017, and completed her M.A. in Economics from MCM DAV College for Women, Panjab University, Chandigarh, in 2019. She was awarded a Ph.D. in Economics from Amity University, Noida, in 2024. Her doctoral research focuses on the socio-economic conditions, efficiency, and sustainability of traditional artisanal sectors, particularly the pottery industry in North India.

Dr. Jaspal Singh is an accomplished economist with over a decade of professional experience in agricultural economics and rural development. He is currently serving as a Consultant in the office of Prof. Ramesh Chand, Hon'ble Member, NITI Aayog. Dr. Singh has previously worked with ICAR-NIAP and as an Assistant Professor in colleges affiliated with Punjabi University. He holds a Ph.D. in Economics and specializes in data analysis, resource use efficiency, and rural employment. A prolific researcher, he has co-authored several academic papers and books and is the Editor of the Indian Journal of Economics and Development. He has been recognized with multiple awards for his contributions to research and policy in the field of agricultural economics.

Anjali Bansal is a development economist with extensive experience in policy research, data analytics, and program evaluation. She is currently a Consultant at KPMG, previously serving in key roles at NITI Aayog, UNDP, and IPE Global. Anjali is pursuing a Ph.D. in Economics from IIT Roorkee, with her research focusing on the future of work, technology, and labor markets. She holds a Master's degree in Economics from Banaras Hindu University (Gold Medalist) and a Bachelor's from the University. Her expertise spans skill development, gender, care economy, and macroeconomic analysis. She has published in reputed journals, co-authored policy reports, and contributed to national development strategies. Her work combines rigorous empirical research with practical policy insights.C-NET in 2019 and completed her internship at NITI Aayog, where she contributed to policy-oriented economic research. Dr. Jakhar has taught as a visiting faculty at Amity University, Noida, and is currently serving as an Assistant Professor at SGT University, Gurugram. Her research has been published in several Scopus and Web of Science-indexed journals, and she has presented papers at national and international conferences on topics ranging from artisanal livelihoods to economic productivity and sustainability. Her core areas of interest include development economics, livelihoods of marginalized communities, production efficiency, and rural enterprise development.

Chapter 12
India's Pursuit of Sustainable Development Goals in Trade

Akhila V. and **Jasdeep Kaur**

Abstract Global sustainability and human well-being are significantly impacted by international trade. According to the 2030 Agenda for Sustainable Development, international trade is acknowledged as a catalyst for inclusive economic growth, poverty reduction, and a crucial tool for attaining the SDGs. The United Nations has embraced 17 Sustainable Development Goals (SDGs) comprising 169 targets. As the world's fastest-growing major economy, India is currently in a unique position to fulfil its commitments to inclusive and sustainable development. Internationally, the country has played a pivotal role in shaping the SDGs and maintaining equilibrium among their three pillars: economic, social, and environmental aspects. India has engaged in trade agreements with numerous nations, aiming to enhance its external trade. The Foreign Trade Policy of India consistently prioritizes significantly expanding the country's global merchandise trade share. Consequently, the Government of India has been implementing various measures to boost trade with other nations, adopting policies and procedures that facilitate the increase of both exports and imports worldwide. In this paper, we will be discussing India's pursuit to achieve sustainable development goals in trade, the current scenario, the measures taken by the Indian government to align its trade practices with sustainable development objectives, and the impacts of international trade on progress towards achieving the SDG targets, challenges, opportunities, future outlook, and suggestions.

The chapters in this book are presented in the "International Conference on Legal and Economic Synergy for an Equitable Society, organised by the University Institute of Laws at Panjab University Regional Centre, Ludhiana, India on February 1st, 2024".

Akhila V. (✉)
KMC College of Law, Tirupur, Tamil Nadu, India
e-mail: akhilavijila001@gmail.com

J. Kaur
District and Session Court, Ludhiana, Punjab, India
e-mail: jaskainth930@gmail.com

A. Virk et al. (eds.), *Law, Economy, and Society: Insights into Legal and Economic Issues in India*, Accounting, Finance, Sustainability, Governance & Fraud: Theory and Application, https://doi.org/10.1007/978-981-95-3361-9_12

Keywords International trade · Sustainable development goals · Economic growth · India · Trade agreements

12.1 Introduction

Global sustainability and human well-being are significantly impacted by international trade. According to the Sustainable Development Agenda of 2030, international trade is recognized as an essential instrument for achieving the SDGs and as a stimulus for economic growth and the reduction of poverty. The seventeen Goals for Sustainable Development have been adopted by the UN, which comprises 169 targets. As the largest and most rapidly developing economy in the world, India currently has a unique opportunity to keep its promises of a comprehensive and resilient future. Additionally, the country has played a crucial role in creating the goal for sustainable development worldwide and balancing it among the three key pillars: Economy, Society, and Environment. India is engaged in trade agreements with numerous nations, aiming to enhance its external trade. The Foreign Trade Policy of India consistently prioritizes significantly expanding the country's share of global merchandise trade. Consequently, the Indian government has been taking several actions to increase trade with other countries, adopting policies and procedures that facilitate the growth of both exports and imports worldwide.

The inclusion of Sustainable Development Goals in trade occurs when trade agreements and practices align with the overarching objectives of the United Nations 2030 Agenda for Sustainable Development. The Sustainable Development Goals are seventeen interconnected goals that address a wide range of problems at both the domestic and international levels, including poverty, inequality, climate change, environmental degradation, and peace. Reintegration of the SDGs into trade means recognizing that trade can also be a force for good in achieving sustainable development.

This paper discusses India's approach to realizing the SDGs in trade. It illustrates the background of the current state of the case; the actions taken by governments in India to enable them to achieve their long-term trade goals, and the implications of global trade on actualizing SDG targets. The paper concludes by offering recommendations for implementation.

12.2 Sustainable Development Goal

Sustainable Development is a holistic concept designed to ensure that current needs are met without causing harm or disadvantage to the capacity of future generations to meet their own needs and requirements. It is about finding a wise balance between economic, social, and environmental factors. Rooted in the belief that development must be enduring and inclusive, sustainable development encompasses a holistic approach to address interconnected challenges facing societies globally. It is a dynamic and evolving concept that provides a roadmap for responsible and inclusive global progress. By embracing the principles of interconnectedness, equity, and environmental stewardship, societies can work towards a future where opulence, welfare, and ecological well-being coexist harmoniously. To achieve the goal of a sustainable and resilient world, cooperation at the local, national, and international levels is crucial.

The pursuit of these development objectives represents a global commitment to addressing pressing challenges and promoting a sustainable future. Adopted in 2015 by every Member State of the United Nations, the goals provide a comprehensive framework for promoting prosperity, equality, environmental sustainability, and social well-being worldwide. The seventeen goals acknowledge the intricate interplay between economic, social, and ecological dimensions. It recognizes that global challenges require collective action and a shared responsibility to build a more inclusive and sustainable world.

The Political Forum at the Highest Level on Sustainable Development, held annually, serves as the principal UN platform for monitoring and assessing progress toward the SDGs. The UN Department of Economic and Social Affairs' (UNDESA) Division for Sustainable Development Goals (DSDG) plays a pivotal role in offering substantial assistance and development of capability for the SDGs and associated thematic areas. These include partnerships, issues concerning States in Development on Small Islands, urbanization, energy, climate change, oceans, water, and transportation. The successful realization of the Goals throughout the UN system, alongside advocacy and communication efforts for the SDGs, heavily relies on the DSDG.

12.3 India and Sustainable Development Goal

With a steadfast commitment to achieving the Sustainable Development Goals (SDGs), India is embarking on a transformative journey to foster inclusive and sustainable development. India's firm resolve to prioritize sustainable development has garnered substantial national and international attention. By harmonizing economic growth with societal advancement and environmental conservation, India strives to ensure a better quality of life for its people while nurturing a greener planet. Emphasizing the belief of "leave no one behind," the Indian government has taken significant strides to integrate the SDGs into its development policies and programs. India has implemented numerous poverty alleviation programs and endeavors which aim to provide sustainable livelihood opportunities, social protection, and access to essential services to the marginalized and vulnerable sections of society, such as the, "Mahatma Gandhi National Rural Employment Guarantee Act" and the "National Social Assistance Program". By recognizing the transformative power of education through such programs, India has embarked on an ambitious mission to enhance the accessibility and quality of education at all levels. Initiatives such as the "Sarva

Shiksha Abhiyan" (SSA) and the National Education Policy (NEP) aim to provide inclusive, equitable, and lifelong learning opportunities for every child and adult, regardless of their socio-economic background.

India's rapid urbanization calls for sustainable urban planning and development, and the Indian government has taken many initiatives on it, such as the Clean India mission, the Rejuvenation Atal Mission, Smart cities, and the Amrut urban transformation. These missions are aimed at fostering sustainable urban infrastructure, efficient waste management, and access to basic sanitation facilities. Recognizing the urgent need to combat climate change and conserve natural resources, India is actively engaged in various climate mitigation and adaptation efforts. The National Solar Mission, the International Solar Alliance, and afforestation projects like the Green India Mission exemplify India's commitment to clean energy transition, sustainable agriculture, and environmental conservation.

India acknowledges the crucial role of gender equality in achieving sustainable development. Through legislative measures, awareness campaigns, and the Beti Bachao Beti Padhao initiative, India strives to eliminate gender-based discrimination, ensuring active participation and empowering women in all spheres of life.

12.4 Sustainable Development Goal in Trade

Trade and sustainable development are addressed in the Johannesburg Plan of Implementation, Chapters V and X, and Chapter 2 of Agenda 21. Both positive and negative effects of trade liberalization and globalization may be felt on sustainable development. It remains essential to support all developing countries in their efforts to join and benefit from the multilateral trading system. The multilateral trading system's contributions must be prioritized to strengthen sustainable development.

The SDG Agenda of 2030 characterizes international commerce as "an engine for inclusive economic growth and poverty reduction, contributing to the promotion of sustainable development". A pledge for promoting "meaningful" liberalization of trade for the following 15 years to increase trade contribution for achieving the 2030 goal was made by the UN States after the ratification of the 2030 Agenda. It is envisaged that trade will facilitate the realization of the SDGs within this framework.

Important organizations that address trade and SDG issues include the UNCTAD, the WTO, and the International Trade Centre. They analyze policies, monitor trends, and train analysts to utilize trade as a catalyst for sustainable progress (Fig. 12.1).

Sustainable Development Goals in Trade:

- **GOAL 17.10**—The SDG Goal of 17.10 states to promote an open, equal trading system without any discrimination or barrier under the World Trade Organization and the Doha Development.
- **GOAL 17.11**—To boost the international sales of nations in the phase of developing and to increase the global proportion of the least developed countries by twofold by the year 2030.

Fig. 12.1 *Sustainable Development Goal for trade*. Source: United Nations Department of Economic and Social Affairs (UNDESA), *Sustainable Development Goals Portal*, https://sdgs.un.org/goals

- **GOAL 17.12**—In agreement with the World Trade Organization to institute a framework wherein markets are accessible without tariffs or trade restrictions or without duties or quotas, and promoting them to take part in global trade by ensuring that preferential Rules of origin are straightforward and uncomplicated for imports originating from the least developed countries.
- **GOAL 2.b**—In accordance with the Doha Development mission to eliminate all forms of subsidies on agricultural export, mitigate barriers to trade, and ensure transparency and accuracy in global agricultural markets.
- **GOAL 3.b**—In alignment with the Doha Declaration regarding the Trade Related Aspects of Intellectual Property Rights to make medicines and vaccines accessible to everyone without any disparity. The goal also affirms the rights of the developing countries regarding Intellectual Property, particularly to safeguard the health of the public by ensuring access to essential medicines.
- **GOAL 8.a**—To support the developing countries by increasing aid for trade and to support the nations with the lowest levels of development through the Enhanced Integrated Framework for trade, a program which aims at bolstering trade capacity and opportunities.
- **GOAL 10.a**—In agreement with the World Trade Organization Agreement to implement Special and Differential Treatment (SDT) measures tailored for the developing countries and countries that are least developed to address the trade challenges.

For the success of the SDG Goals, it demands collective work and comprehensive collaboration. Countries must work together to establish an enabling environment for trade by addressing policy gaps, strengthening institutions, and promoting transparency. Recognizing trade as a catalyst for sustainable development, cooperation among all the institutions is essential. Collective efforts facilitate knowledge-sharing, technology transfer, and innovative solutions to overcome challenges. By integrating SDGs into trade, nations aim to achieve the potential of international commerce to achieve eco-friendly and inclusive growth, creating a balance between economic growth, social progress, and environmental stewardship. This approach reflects a commitment to a more holistic and responsible approach to global trade. Key aspects of SDG integration in trade include:

- **Inclusive Economic Growth:** Encouraging trade policies that foster inclusive economic growth by providing opportunities for all societal sectors, including marginalized groups and small businesses.
- **Poverty Alleviation**: Utilizing trade as a tool for poverty reduction by creating employment opportunities, enhancing income levels, and supporting economic empowerment in developing regions.
- **Environmental Sustainability**: Ensuring that trade practices are environmentally sustainable, minimizing negative impacts such as pollution, resource depletion, and deforestation. This may involve promoting eco-friendly production methods and sustainable supply chains.
- **Gender Equality**: Integrating gender-responsive trade policies that address disparities and promote equality in terms of access to trade opportunities, resources, and benefits for both genders.
- **Quality Education and Skill Enhancement:** Aligning trade policies with goals related to education and skill development to enhance human capital, ensuring that individuals are equipped with the knowledge and skills required for meaningful participation in the global economy.
- **Infrastructure Development**: Using trade as a catalyst for infrastructure development, especially in developing countries, to improve connectivity, transportation, and logistics, thereby facilitating smoother trade flows.
- **Global Partnership:** Fostering international cooperation and partnerships to help the developing countries to integrate in the global economy and to address common challenges related to trade, including policy coherence and fair trade practices.
- **Ethical consumption and sustainable Production:** Encouraging responsible utilization and manufacturing patterns by advocating sustainable & ethical practices throughout the supply chain.
- **Climate Action**: Integrating climate-friendly trade policies that encourage the adoption of energy sources, which would aid in reducing carbon emissions and mitigating climate change.
- **Peace, Justice, and Strong Institutions:** Promoting fair trade practices, combating corruption, and enhancing the rule of law in the trade sector to ensure that trade activities contribute to peace, justice, and the strengthening of institutions.

12.5 India's Focus on Sustainable Development: Goals in Trade

From Histories, Trade is shown as a powerful tool for poverty eradication, which aids in the nation's prosperity. The Millennium Development Goal to cut poverty in half by 2015 was achieved early because of the exceptional decline in poverty levels brought about by rapid growth. Trade benefits emerging nations because it enables them to effectively use their resources by specializing in the manufacturing of the products and services that are in demand in the global economy or competitive markets. Trade also boosts economic growth in each of these nations. Trade can boost growth and provide the resources required to carry out additional development goals in the societal and environmental domains. By creating new job opportunities, trade directly helps in reducing the nation's poverty rate.

India, as a signatory to the SDG 2030 Agenda, has dedicated itself to integrating Goals on sustainable development (SDGs) into various aspects of its policies, including trade. This step reflects a multifaceted effort to advance social, environmental, and economic objectives in the quest of a sustainable and inclusive future. The 2030 Agenda holds the objectives aimed at fostering sustainable development, and the Indian government is fully committed to achieving these goals. The SDGs and India's national development goal, "Collective Efforts, Inclusive Growth," also known as "Sabka Saath, Sabka Vikas," go hand in hand.

12.6 Economic Growth and Decent Work (SDG 8)

By facilitating international commerce, India's trade policies seek to advance economic growth. An attempt is made to contribute to SDG 8 targets on sustained and inclusive economic growth by streamlining processes, lowering obstacles, and improving the ease of doing business. Trade activities boost economic sectors, resulting in the creation of jobs and chances for decent work, which is consistent with the goal of SDG 8 on full and productive employment.

India aims to have every citizen, male or female, including those with impairments, employed decently and contributing to the national GDP by the year 2030. The Indian government has launched a number of initiatives by improving skill enhancement, creating more job opportunities, and by working on the economic growth for the welfare of the general public to succeed in achieving these goals.

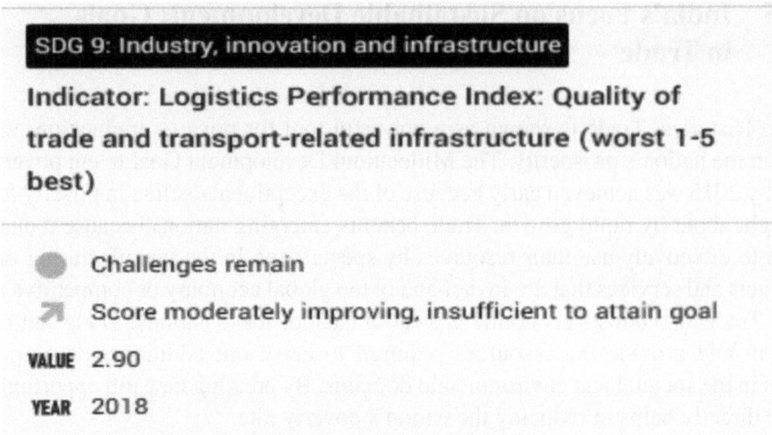

Fig. 12.2 SDG 9 industry, innovation and infrastructure

12.7 Innovation in Trade Practices (SDG 9)

Industry, Innovation, and Infrastructure India's trade policy places a strong emphasis on industry innovation, which is consistent with SDG 9's objectives of supporting equitable and sustainable industrialization and building resilient infrastructure. Infrastructure Development: In line with SDG 9's emphasis on creating robust infrastructure, trade-related infrastructure initiatives improve connectivity, logistics, and transportation. Presently, India is at a value of 2.9, and the aim of India is to reach Aim of India is to reach a value of 3.8 which if the country continues its performance, they may succeed in reaching the goal (Figs. 12.2 and 12.3).

To strengthen the infrastructure, the Indian Government has taken many initiatives, such as the Pradhan Mantri Sadak Yojana, Sagarmala, Bharatmala, etc., and for the development of innovation and industry sector projects, such as Digital India and Made in India, has been successfully implemented.

12.8 Sustainable Trade Practices: Ethical Production and Consumption (SDG 12)

India encourages sustainable trade practices to promote responsible production and consumption. This entails reducing waste, encouraging environmentally friendly production techniques, and implementing the ideas of the circular economy. Promoting and upholding eco-certifications in commerce helps to fulfill SDG 12's objectives regarding sustainable patterns of production and consumption. The LIFE or Lifestyle for Environment Mission was launched by the Indian Government to encourage individual and collective action for safeguarding the environment.

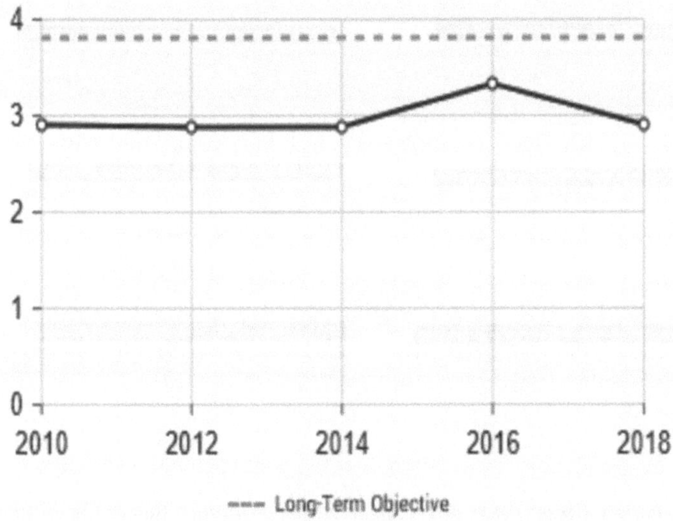

Fig. 12.3 *Trade and SDG Linkages Framework.* Source: World Trade Organization (2018), *Mainstreaming Trade to Attain the Sustainable Development Goals*, Geneva

Analysis conducted by the International Energy Agency found that, to reduce annual global carbon dioxide emissions by around 2 million tons by the year 2030, global adoption of LIFE-targeted measures like eco-friendly choices and behavioral modifications must be promoted (Fig. 12.4).

12.9 Climate Action: Green Trade Initiatives (SDG 13)

India supports green trade initiatives and recognizes the significance of trade in addressing climate change. This entails encouraging environmentally friendly behaviors, cutting carbon emissions from supply chains, and taking part in global initiatives to tackle climate change. Trade in Renewable Energy: Encouraging trade in renewable energy technologies and services is consistent with SDG 13's goal of accelerating the fight against climate change and its effects. The "One Sun One World One Grid" (OSOWOG) action plan was implemented by the International Solar Alliance, which hopes to construct a solar grid shared at an international level. The project calls for the swift and widespread distribution of solar energy, stabilization of the energy supply, promotion of study and development to address changes in sunshine availability, and constant maintenance of dependable base load capacities.

The International Energy Agency's recent report, "World Energy Investment 2023," emphasizes that investing in green energy has surged, mostly driven by renewable energy and electric vehicles (EVs). According to the report, India has

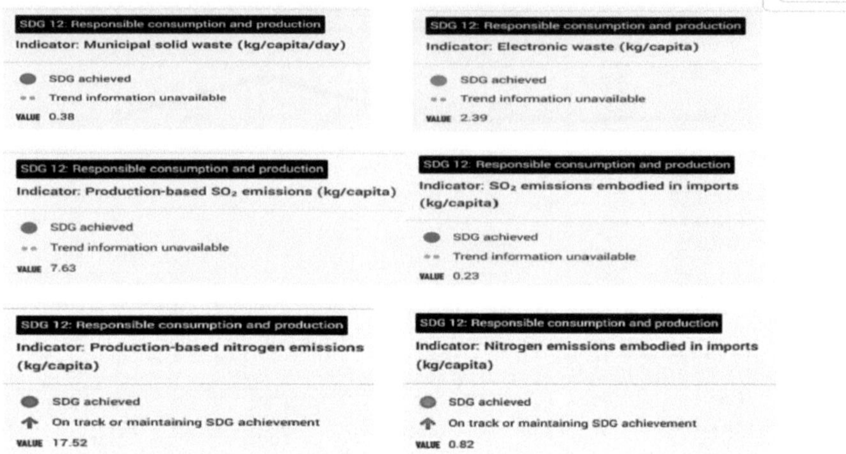

Fig. 12.4 *India's Green Trade and Climate Action Measures.* Source: International Energy Agency (2023), *World Energy Investment Report*, Paris

demonstrated leadership in promoting energy transition through a range of initiatives implemented in alignment with its net-zero goals.

12.10 Gender Equality: Inclusive Trade Policies (SDG 5)

India seeks to advance inclusion through trade policies in order to advance gender equality. In order for women to profit from the economic opportunities created by trade, efforts are undertaken to guarantee that they actively participate in trade-related activities.

To achieve the SDG goal 5 of Gender Equality & Inclusive Trade Policies, schemes and initiatives to support women entrepreneurs by increasing their access to education, skill enhancements, and institutional credit should be promoted on a global level. Women are becoming more economically empowered as a result of this shift from women-led development to women's development, which will also raise their share of the GDP of the country. It is anticipated that India's GDP might increase by US$770 billion by 2025 if it provides equal opportunity to women.

The Atmanirbhar Bharat Development scheme launched by India focuses mainly on women's empowerment. Other policies and programs initiated by India at the national level have pledged to promote the complete improvement of women in all phases. The core of India's G20 Presidency is also the country's model of women-led development.

12.11 Partnership for the Goals (SDG 17)

India actively participates in international trade cooperation for sustainable development. This includes commercial agreements, both bilateral and multilateral, that support the objectives of SDG 17, which are to develop implementation mechanisms and rekindle international cooperation. Knowledge-sharing, capacity-building, and technology transfer are all components of collaborative trade initiatives that support SDG 17's objectives of strengthening international cooperation.

The policy alliances offer a setting and a stage for interdisciplinary cooperation. Along with tackling energy security (SDG 7), food security and poverty (SDGs 1 and 2), fostering innovation and transformation in technology (SDG 9), and supporting climate action (SDG 13), these collaborations also address other important concerns.

Through its policies and actions, India has greatly profited from both bilateral and multilateral international partnerships as well as the global common good. This reinforces the central tenet of Indian foreign policy, "Vasudhaiva Kutumbakam" (the world is one family). The 'Lifestyle for the Environment (LiFE) Movement', the International Solar Alliance (ISA), the India Stacks initiative for Digital Transformation, and the Coalition for Disaster Resilient Infrastructure (CDRI) are a few instances in this regard.

India is not only a developing country but has also used its firsthand experience to create international development partnerships with a development plan that is people-centered and sustainable (Fig. 12.5).

At present, India is a signatory to 13 regional trade agreements (RTAs) and free trade agreements (FTAs) with other nations. It also has six PTAs (preferential trade agreements) with limited reach. Government agencies, trade associations, and industry groups are represented on the FTA Monitoring Committee, which has been

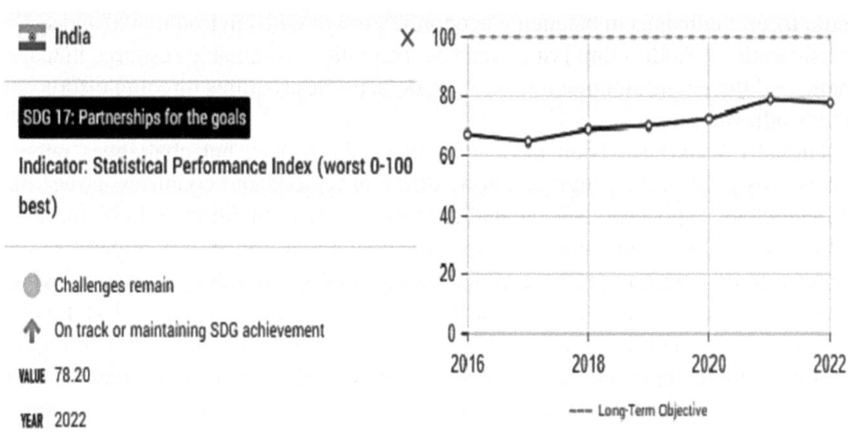

Fig. 12.5 *India's Partnerships and Global Cooperation for SDGs*. Source: Indian Council of World Affairs (ICWA), *SDG 17 Partnership for Goals – India's Approach and Initiatives*

established to investigate problems related to the abuse of FTA provisions and provide solutions. For the benefit of all participating nations, the Indian government has been a part of and developed major regional and international economic alliances for nearly ten years.

12.12 Well-being and Good Health (SDGs 3)

Pharmaceutical Exports: India's pharmaceutical sector is vital to the world's supply of reasonably priced medications, which helps achieve SDG 3's goals of promoting health and well-being.

India has been making great efforts to address the goal three's requirements to improve the nation's healthcare system and to achieve those some of the initiatives started by the government are, the Ayushman Bharat, "the National Programme for Prevention and Control of Cancer, Diabetes, Cardiovascular Diseases, and Stroke (NPCDCS)", "National Health Mission", "National Mental Health Programme (NMHP)", "National Rural Health Mission (NRHM)", "National Urban Health Mission (NUHM)", the National Programme for Control of Blindness, the National Leprosy Eradication Programme, the National Mental Health Yojana (PMJAY), the National Programme for Control of Blindness, the National Programme for Prevention and Control of Cancer, Diabetes, and Stroke, etc. India aims to provide affordable and accessible healthcare for all regardless of socio-economic background, geographic location or any other barriers.

12.13 Progress and Challenges

India faces challenges in balancing economic growth with environmental and social considerations. Addressing issues such as inequality, sustainable resource management, and the environmental impact of trade activities requires ongoing efforts and policy adjustments.

Global efforts have been made to achieve the SDGs, but challenges persist. Disparities exist in the progress among different regions and countries. Issues like climate change, poverty, and inequality require urgent attention and collaborative solutions. For the sustainable development agenda to advance, it is imperative that an open trade environment be established, especially through e-commerce, which ensures a secure environment for businesses and consumers while fostering equal prospects for economic growth and development. Striking this balance will require an international approach that promotes dialogue and the open exchange of information between actors to comprehend connections among trade, economic progress, and the reduction of poverty, as well as to address the challenges posed by a constantly changing trade and economic landscape.

12.14 Conclusion

The Goals of Sustainable Development serve as a roadmap for a more equitable, eco-friendly, and prosperous world. Achieving these goals demands sustained commitment, innovative solutions, and global cooperation. As we work towards 2030, the international community must remain dedicated to the principles of sustainability, inclusivity, and shared responsibility for the betterment of present and future generations. India's steadfast commitment to the Sustainable Development Goals has positioned it as a global leader in sustainable development. India works to prosper and gain a sustainable future for all its citizens by integrating these goals into policy frameworks and implementing transformative initiatives. India strives to pave the way towards a thriving and sustainable future for all its citizens.

As India continues to navigate the complexities of global trade, its commitment to integrating sustainable development principles serves as a positive catalyst for addressing not only domestic challenges but also contributing to the broader global agenda. However, ongoing efforts, adaptability, and continuous evaluation are essential to make sure that trade policies effectively contribute to the realization of SDGs and a more sustainable and equitable future.

References

Helble M, Shepherd B, Asian Development Bank Institute (2017). https://www.adb.org/sites/default/files/publication/327451/adbi-win-win-how-international-trade-can-help-meet-sdgs.pdf#page=20. Accessed 15 Mar 2025

Icwa (n.d.) SDG 17 Partnership for Goals- India's Approach and Initiatives - Indian Council of World Affairs (Government of India). https://www.icwa.in/show_content.php?lang=1&level=3&ls_id=9773&lid=6254

IEA (2023) LiFE lessons from India. IEA, Paris. https://www.iea.org/reports/life-lessons-from-india, License: CC BY 4.0

LiFE lessons from India–Analysis-IEA (n.d.) IEA. https://www.iea.org/reports/life-lessons-from-india

Localising SDGs Early Lessons From 2019, NITI AYOG (2019)

Mainstreaming trade to attain the sustainable development goals (2018) https://doi.org/10.30875/9c96f135-en

Mainstreaming trade to attain the Sustainable Development Goals, WORLD Trade Organization. https://www.wto.org/english/res_e/booksp_e/sdg_e.pdf

Nina E, Nina E (2023, June 7) The role of international trade in achieving the UN's sustainable Development Goals - International Trade Council. *International Trade Council - Solving trade-related issues*. https://tradecouncil.org/the-role-of-international-trade-in-achieving-the-uns-sustainable-development-goals/

Perappadan BS (2023, April 26) India intends to make healthcare affordable for whole world: PM Modi. The Hindu. https://www.thehindu.com/news/national/indias-goal-is-to-make-healthcare-accessible-and-affordable-not-only-for-our-citizens-but-for-the-whole-world-pm-modi/article66781569.ece

Roadmap for Women's Economic Empowerment in India | IBEF (n.d.) India Brand Equity Foundation. https://www.ibef.org/blogs/roadmap-for-women-s-economic-empowerment-in-india
Sachs JD, Lafortune G, Fuller G, Iablonovski G (2023) Sustainable Development Report 2023: Data Explorer [Statistical Performance Index (SDG 17) – India, 2016–2022 graph]. UN Sustainable Development Solutions Network. Retrieved February 2024, https://dashboards.sdgindex.org. Accessed 15 Mar 2025
SDG Transformation Center's data hub (n.d.-a). https://datahub.sdgtransformationcenter.org/reports/sustainable-development-report/indicators/sdg9_lpi/ratings
SDG Transformation Center's data hub (n.d.-b). https://datahub.sdgtransformationcenter.org/reports/sustainable-development-report/goals/17/ratings
SDG Transformation Center's data hub (n.d.-c). https://datahub.sdgtransformationcenter.org/reports/sustainable-development-report/goals/12/ratings
Sustainable Development Goals (n.d.-a). https://education.nationalgeographic.org/resource/sustainable-development-goals/
Sustainable Development Goals (n.d.-b) UNDP. https://www.undp.org/sustainable-development-goals
THE 17 Goals | Sustainable Development (n.d.). https://sdgs.un.org/goals
Trade | Department of Economic and Social Affairs (n.d.). https://sdgs.un.org/topics/trade
Trading Into Sustainable Development: Trade, market access, and the sustainable Development goals, UNCTAD/DITC/TAB/2015/3, United Nations Publication, ISSN 1817-1214
Transforming our world: the 2030 Agenda for Sustainable Development | Department of Economic and Social Affairs (n.d.). https://sdgs.un.org/2030agenda
UN (2019) Localising SDGs early lessons from India 2019, NITI Aayog. https://www.niti.gov.in/sites/default/files/2020-07/LSDGs_July_8_Web.pdf. Accessed 15 Mar 2025
UN, Global Sustainable Development Report (GSDR) 2023 | Department of Economic and Social Affairs, https://sdgs.un.org/gsdr/gsdr2023. Accessed 15 Mar 2025
Voluntary National Review Report on Implementation of Sustainable Development Goals United Nations High Level Political Forum 2017, The high-level political forum on sustainable Development, New York/July 2017, https://sustainabledevelopment.un.org/content/documents/16693India.pdf
Win–Win: How international trade can help meet the sustainable development goals. In: Helble M, Shepherd B (eds) Asian Development Bank Institute
World Energy Investment 2023–Analysis-IEA (n.d.) IEA. https://www.iea.org/reports/world-energy-investment-2023

Akhila V. obtained her B.A.LL.B. (Hons.) degree from Lovely Professional University, Jalandhar, in 2021, and completed her Master of Laws (LL.M.) in 2022 from the same university. She is enrolled as an Advocate with the Bar Council and is currently working as an Assistant Professor of Law at KMC Institute of Law, Tirupur. She has previously served in similar roles at GHG Institute of Law, Ludhiana, and CT Institute of Law, Jalandhar. Her academic and research interests focus on intellectual property rights, environmental law, digital labour platforms, and the rights of marginalized communities. She holds a Certificate in Human Rights from IGNOU. She has participated in an international training program on Digital Labour Platform Policies and Case Law, conducted by the ITC-ILO and co-financed by the Italian Ministry of Foreign Affairs and International Cooperation. Akhila has presented and published over ten research papers at national and international forums on themes such as outcome-based education, labour law reforms, domestic violence during the COVID-19 pandemic, and global citizenship education. She has also contributed as a resource person and has actively coordinated legal, academic, and cultural events across institutions.

Jasdeep Kaur is a dedicated legal professional and academician, currently practicing as an Advocate at the District and Sessions Court, Ludhiana. She obtained her B.A.LL.B. (Hons.) degree from Panjab University, Chandigarh, in 2020, and went on to complete her Master of Laws (LL.M.) from Kurukshetra University in 2022. Jasdeep is enrolled with the Bar Council and has handled civil, criminal, and service law matters with a focus on legal drafting, case analysis, and courtroom advocacy. In addition to her legal practice, she has served as an Assistant Professor at the GHG Institute of Law, Ludhiana, where she taught core law subjects, including Criminal Law, Business Law, Labour Law, Constitutional Law, and Intellectual Property Rights. During her academic tenure, she actively mentored students and organized academic, legal, and cultural events. She has also contributed as a resource person and coordinator for various institutional initiatives aimed at legal awareness and student development. Jasdeep's academic and research interests include intellectual property rights, humanitarian law, labour law reforms, National Education Policy, women's empowerment, and the rights of marginalized communities. She holds a Certificate in Human Rights from IGNOU and has presented and published over eight research papers at national and international platforms. Her work has addressed critical issues such as the rights of the LGBTQ+ community, labour law reforms, artificial intelligence and IPRs, One Nation One Election, and human rights.

Jasdeep Kaur is a dedicated legal professional and academician, currently practicing as an Advocate at the District and Sessions Court, Ludhiana. She obtained her B.A.LL.B. (Hons.) degree [illegible]

Chapter 13
Law and Economics of Oil Spills in India: A Critical Analysis

R. Vijaya Bharathi and S. R. Ramya

Abstract Oil spills are known to be one of the most hazardous industrial accidents. The effects of these oil spills are tenacious and cause severe environmental damage. These oil spills involve enormous amounts of money and manpower for removal and restoration. The fixation of liability becomes difficult if the cause of the oil spill is based on a natural disaster. The international instruments include the International Convention for the Prevention of Pollution from Ships, 1973, which governs accidental and operational spills, the International Convention on Civil Liability for Oil Pollution, 1969, which fixes the liability and the International Convention on the Establishment of an International Fund for Oil Pollution Damage, 1971, which financially assist in mitigating the environmental costs. The conventions are given enforceability by incorporating them in India through the Merchant Shipping Act, 1958. Despite several legislative and administrative frameworks, it is the locals and residents who end up removing the oil spills instead of the polluters or government agencies. Most of the time, the cleaning is done without the aid of any appropriate gear or technological equipment, which raises concerns about health and hygiene. This research attempts to critically analyze the legal and economic aspects of oil spills in India through the case study method. It also aims to address the issues in the legal framework and bridge the gap between the lacunae in the legislative framework and the environmental costs and concerns involved in oil spills in India.

The chapters in this book are presented in the "International Conference on Legal and Economic Synergy for an Equitable Society, organised by the University Institute of Laws at Panjab University Regional Centre, Ludhiana, India on February 1st, 2024".

R. Vijaya Bharathi (✉)
VIT School of Law, Vellore Institute of Technology (VIT), Chennai, India
e-mail: vijayabharathi.r@vit.ac.in

S. R. Ramya
The TamilNadu Dr. Ambedkar Law University, Chennai, India
e-mail: tazettaprimrose@gmail.com

A. Virk et al. (eds.), *Law, Economy, and Society: Insights into Legal and Economic Issues in India*, Accounting, Finance, Sustainability, Governance & Fraud: Theory and Application, https://doi.org/10.1007/978-981-95-3361-9_13

Keywords Oil spills · Cost · MARPOL · Liability · Damage

13.1 Introduction

With the increasing growth in international trade and globalisation, the use of marine transport has increased manyfold. There is also a considerable increase in offshore activities. But these developments come with a price. In the name of development, in several instances, the environment has been jeopardized. Most marine pollution is caused by human activities, including emissions and oil spills. The issue with such pollution is that the effects are neither limited by time nor by geographical limits. The widespread usage of marine transportation as part of trade and commerce also poses one such distress, which is oil spills.

The Supreme Court of India in *M. V. Elisabeth v. Harwan Investment And Trading* stated that "India is lagging behind compared to various other countries in ratifying the various international conventions as well as adopting the beneficial provisions of the conventions that intend to facilitate international trade and free movement. Although these conventions have not been adopted by the legislative framework in India, these principles incorporated in the conventions are derived from the common law of nations and can be seen as those embodying the felt necessities of international trade. As such, these principles are part of the common law of India and applicable for the enforcement of maritime claims against foreign ships". (M.V. Elisabeth v. Harwan Investment and Trading 1993)

Oil spills represent a complex intersection of environmental, economic, and legal challenges, with profound implications for affected ecosystems, communities, and industries. In the context of India, a country with a rich maritime heritage and a rapidly growing energy sector, the occurrence of oil spills poses significant concerns for environmental sustainability, economic development, and legal governance. This paper undertakes a critical analysis of the law and economics surrounding oil spills in India, aiming to shed light on the multifaceted issues at play and explore potential avenues for improvement.

The economic dimensions of oil spills in India are substantial, encompassing direct costs associated with spill response, cleanup operations, and compensation payments, as well as indirect costs stemming from damage to marine ecosystems, loss of livelihoods for coastal communities, and impacts on tourism and fisheries industries. Moreover, the long-term environmental repercussions of oil spills can exert lasting economic effects, undermining ecosystem services, hindering coastal development, and eroding investor confidence in maritime industries.

At the heart of addressing oil spills in India lies the legal framework governing spill prevention, preparedness, response, and liability. Existing laws and regulations, including the Merchant Shipping Act, 1958, the Environment (Protection) Act, 1986, and the National Oil Spill Disaster Contingency Plan (NOSDCP), provide a foundation for addressing oil spill incidents. However, gaps and challenges persist in the implementation and enforcement of these regulations, raising questions about their effectiveness in ensuring comprehensive protection of marine environments and accountability for spill-related damages.

This critical analysis seeks to interrogate the strengths and weaknesses of the current legal and regulatory regime governing oil spills in India. By examining key case studies, policy debates, and economic assessments, we aim to identify areas for improvement and innovation in both law and policy. Moreover, we will explore the potential role of financial instruments, such as liability regimes, enhancing preparedness, and facilitating timely and equitable compensation for affected parties.

Ultimately, this paper endeavors to contribute to a deeper understanding of the complex dynamics surrounding oil spills in India, offering insights into how the law and economics can be leveraged to promote sustainable and resilient marine ecosystems, foster responsible resource management, and safeguard the interests of present and future generations.

13.2 Unveiling the Ripple Effects of Oil Spills

The effects of oil spills can be traced across multiple spectrums, ranging from the environment and marine life to the economy. The oil spill can also cause severe damage to marine life, including various species of fish, birds, and other aquatic animals, which may lead to large-scale migration and extinction. The oil droplets on the layer of water will entrap the living organisms of the ocean, including the sea birds, and kill them either by poisoning or hypothermia. Other long-term effects could be on the vulnerable mangroves or the coral reefs (Jernelöv 2010) that are damaged because of the toxicity. The impact on the environment can be long-lasting even after an effective clean-up activity. Though the surface looks clean, it can remain in the environment for quite a long and create harmful sediments. The plants in and around the affected area absorb oil that ultimately damages the plant's growth by making it inappropriate for the environment.

The effects from an economic perspective are also multilateral. Firstly, let us examine from the perspective of loss of energy resources. Loss of large quantities of oil will decrease the availability of its by-products like fuel, which in turn results in a shortage of supply and thereby increases the price of the oil and its by-products. Secondly, there is also an enormous amount of energy, technology, human resources, and financial resources involved in cleaning the spill. Thirdly, it affects the fishing community to a greater extent because, post the oil spill, fishing activities become impossible until the spill is cleaned up. It also has an adverse effect on tourism in the affected area, and because of entry barriers and other restrictions, even the locals would not visit the affected area, thereby putting the local vendors' livelihood in jeopardy.

This is why it is essential that when dealing with sustainability in oil exploration, there must be a balance between the financial returns and the technology transfer in the economic sense. Oil exploration must be such that there is an improved quality of living. Not only economical but environmentally sustainable technology is required to prevent oil spillage, such that the local environment (Pearce and Turner 1990) can be preserved and stress-free.

13.3 Oil Spills: Charting the Course of International Legal Responses

With the international community involved in this transnational event, the applicability of the International norms and Conventions to govern any environmental hazard caused by the spillage of oil becomes important. The activities that affect the other jurisdictions can be governed by the Stockholm Declaration or the Law of the Sea Convention so as to prevent Transboundary Pollution (Olawuyi 2013). Hence, it becomes an international obligation of all parties involved to mitigate the cross-border effects of oil spills. The ones happening on the High Seas become an international concern, while the ones occurring in the vicinity of a territory invoke sovereign issues.

The International Convention for Prevention of Pollution of Sea by Oil, 1954 (OILPOL) recognised that the pollution from oil spill is mostly because of operational discharge while cleaning the oil tanks of the ships. It prohibits such discharge of oil closer to the land or in special areas designated so. But dealing with accidental spills like that of MSC Chitra in the Arabian Sea, offshore instalments, and pipelines needs a different approach. The issue cannot be resolved by the actions of one nation as it is beyond its scope and hence needs an international commitment. It can be termed as a 'collective action problem' and requires a comprehensive legal regime to regulate the movement of oil in the marine environment. The International Convention for the Prevention of Pollution from Ships (MARPOL), 1973 and as modified in 1978 along with OILPOL, deals with the structure and safety standards set for the ships, especially the design of the oil tanker and the operating procedures so as to minimise the oil spill, be it operational or accidental (Suleman 2011). In order to meet international standards and bring about the exchange of information between the states, India also signed the International Convention on Oil Pollution Preparedness, Response and Co-operation (OPRC). Again, the OPRC, like the other environmental conventions, advocates the 'polluters pay' and 'precautionary' principles.

Despite entering into numerous conventions, all propagate civil liability and compensation by providing procedures for the assessment of the quantum and setting up funds or insurance. The principle adopted to govern the liability regime from oil pollution is Strict liability that imposes liability without fault, as applied in *Ryland v. Fletcher,* (Ryland v. Fletcher 1868). The International Convention on Civil Liability for Oil Pollution Damage (CLC), 1969, and the International Convention on the Establishment of an International Fund for Oil Pollution Damage (Fund Convention), 1971, set up compulsory liability insurance for the ship owners and a compensation fund in excess of the owner's liability respectively. There is also one International Convention on Civil Liability for Bunker Oil Pollution Damage, 2001, that specifically deals with liability from spillage of bunker oil that also follows strict liability.

Looking at the Liability and Compensation regime, the first and foremost daunting task is to compute the economic and environmental costs of the oil spill. While the authors can suggest a few criteria to be considered for the computation of such costs and corresponding liability, the economic valuation of the impact of an oil spill that transpires into every aspect of life, present and future, is far-fetched. The victims of an oil spill can range from fishermen to those running a tourism business (Augusyniak 1993). Apart from the flora and the fauna, entire ecosystems can be destroyed, and the resultant changes could have an effect on the future marine systems.

If the difficulties in computing the compensation are of a varying nature, the clean-up costs and the diverse issues related to the process are even more massive. Despite the 'polluters pay' principle, the costs of preventing a spill and reparation of the oil pollution damage, at times, need to be internalised. As such, when it comes to oil production and exploration, International Oil companies (Kuokkanen 2002) bear more risk of fulfilling the obligations and paying the compensation costs in transboundary pollution cases.

The Pollution caused by oil spills may be mitigated by either governmental regulations or the companies under the banner of Corporate Social Responsibility (CSR). The extent to which these voluntary initiatives are effective (Frynas 2012) then depends on the ethics of the companies as it relates to accountability for sustainability. Voluntary practices substantially improve the environment as they also result in higher profits owing to self-reporting. As part of their CSR, if the oil companies themselves can provide an extensive report, then there could be a reduction in the oil spills ascribed to improved conditions of the oil tankers and other ships, causing marine pollution due to oil spills.

Voluntary self-regulation originally began after the two international agreements—the Tanker Owners' Voluntary Agreement concerning Liability for Oil Pollution (referred to as TOVALOP) and the Offshore Pollution Liability Agreement (referred to as OPOL). The Full Cost principle (Sangal 1995) is one such legal norm that, if applied, would make any product that causes environmental damage expensive, be it your everyday automobiles or the commercial vessels that pollute the environment. The Full Cost principle requires the user of the environment to pay the costs of the economic losses that can be directly attributed to the product, as any remoteness in fixing the costs can be seen as overbearing. This principle, to an extent, is implemented via the TOVALOP, 1969, and the Contract Regarding a Supplement to Tanker Liability for Oil Pollution (CRISTAL), 1971. Though these voluntary agreements have been developed under the auspice of the International Maritime Organization, the fact remains that the parties may choose not to participate in these (De la Rue 1997). If they agree to be a part, then the parties may fall under contractual obligations to meet and act according to the terms of the agreements. Then again, these compensation mechanisms only deal with persistent oil that is not defined. From common parlance, persistent oil may be taken as crude oil, heavy diesel oil, fuel oil, and lubricating oil. Further, the oil companies also bear the costs of cleaning up by creating a fund under these voluntary agreements.

Despite these self-regulating mechanisms, the effective contribution to the prevention of oil spills can be seen to be meager through these initiatives, as self-reporting of environmental affairs is not the same as taking preventive measures. The CLC and Fund Convention did extract the liability and compensation fund models from these voluntary agreements and took over the regime of fixing liability and assessing compensation to the affected parties from the TOVALOP and CRISTAL after they expired in 1997 with the international community failing to see the need to renew them after a fulfilled purpose and baton handed over to the International Conventions.

13.4 Navigating the Legal Waves: Oil Spills and the Indian Regulatory Framework

India, being a close aide in the development of the International Maritime Organization Conventions and their implementation, has incorporated many of these International conventions into its regulatory framework (DGS, n.d.). Even after India acceded to the Bunker Convention, the maximum amount of compensation or liability of the polluter for bunker pollution damage computed under the national or international regime is subject to the Convention on Limitation of Liability for Maritime Claims (LLMC) 1976. This is because Article 3 of the LLMC lists those maritime claims under CLC as an exception to the limitation (International Maritime Organization, n.d.) clause, and the Bunker Convention is yet to be incorporated into the Indian legal framework through necessary amendments. As such, governmental authorities find it difficult to recover the costs involved in the prevention and clean-up of these kinds of oil pollution apart from those caused by tankers (Mittal 2023).

The important legal framework of the Merchant Shipping Act of 1958 (MSA) primarily focuses on the prevention and control of marine oil pollution. The MSA works alongside its various rules, including The Merchant Shipping (International Fund for Compensation for Oil Pollution Damage) Rules, 2008, and The Merchant Shipping (Prevention of Pollution by Oil from Ships) Rules, 2010. The CLC has been incorporated along with LLMC in Part XA of the Act, wherein section 352A clearly states that any limitation in the liability will not be extended to oil spills, and Part XB (sections 352G to 352R) deals specifically with the Civil Liability for the Damage caused by Oil Pollution (Merchant Shipping Act, 1958, § 10). This part clearly fixed the Liability on the owner for pollution damage caused by persistent oil for his personal act, omission, negligence, recklessness, etc. However, there is a limitation on liability if he proves the pollution was caused by war situations or the wrongful acts of other persons. Further, it also provides for setting up of a limitation fund and compulsory insurance. Part XC of the MSA (sections 352S to 352ZA) is in line with the Fund Convention in constituting a Compensation fund to which anyone who imports/receives oil in India is to contribute. High courts are given the jurisdiction for such an actionable claim. Apart from these provisions, Part XIA (sections 356A to 356O) deals with general prevention and control of oil pollution. Again, only oil tankers are recognised, and only the process of obtaining their pollution prevention certificates and standards of safety equipment are mentioned, along with the powers of the Central Government to take measures (Merchant Shipping Act 1958, § 365K).

Since the Indian Coast Guard is the coordinating agency for marine pollution (Coast Guard Act 1978, § 14) under the Coast Guard Act 1978, the National Oil Spill—Disaster Contingency Plan (NOSDCP) has been put in place to incorporate the international best practices for the prevention of oil spill and to immediately respond in case of accidents (Rathsharma 2021). The NOSDCP lacks a statutory basis and lack of cohesiveness between the different agencies, and multiple regulations are also adding up to the event of underplaying the extent of damage caused by oil spills.

The NOSDCP, which came into operation in 1996, coordinates the functionalities of different agencies, those of the Coast Guard and the Emergency response services of the state in combating marine pollution. It implements a National Contingency Plan and directs the incorporation of a State Contingency Plan for the surveillance of maritime zones for oil spills and controlling such disasters. It recognizes Oil spills occurring from accidents or illegal operational discharges and tries to facilitate appropriate, rapid, and effective responses to the pollution in any part of the sea that can potentially impact the maritime zones of India.

The National Disaster Management Authority and Crisis Management Groups provide expert guidance in handling major oil spills. Any pollution response is in the order of preventing, minimizing, and mitigating. So is an oil spill contingency plan, with the help of various agencies and support groups. Overseas assistance may also be obtained through invoking OPRC.

The NOSDCP, in its clean-up option, does include dispersant spraying operations, cargo transfer, and mechanical recovery operations to minimize the environmental as well as socio-economical impact. A methodology of oil spill response has to be carefully selected, keeping in mind the natural clean-up process, and this is termed as Net Environmental Benefit Analysis (NEBA) based on the type of spill, the distance, and the weather conditions, wherein the responsibility of the clean-up lies with the port authority.

The Fig. 13.1 depicts the organisations involved in the emergency response to an oil spill disaster as given out by the Indian Coast Guard (NOSDCP, Ministry of Defence, Government of India 2015).

The 8th National Level Pollution Response Exercise conducted in April 2022 by the Indian Coast Guard, with stakeholders from international organisations and around 20 countries, showcased India's prowess in responding to oil spills. The NOSDCP is validated not only at the national but also the regional level by this workshop exercise under the aegis of the South Asian Cooperative for Environment Protection (Press Information Bureau 2022).

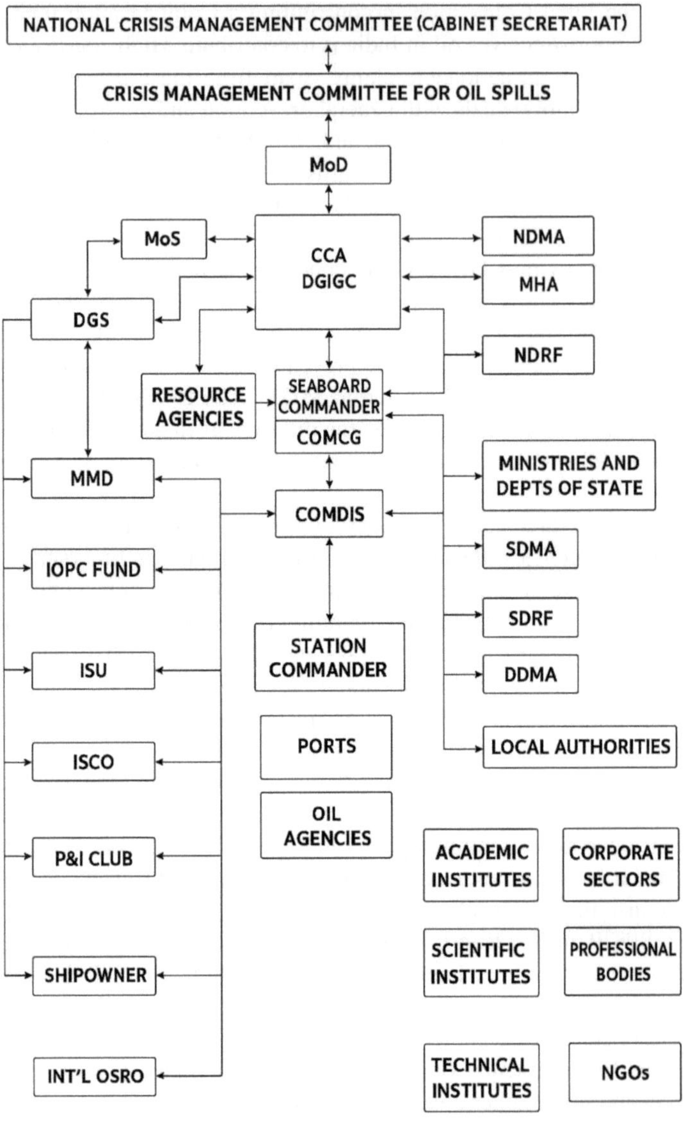

Fig. 13.1 Organisation for emergency response for oil and chemical spill disasters. Source: NOSDCP Plan Manual, Ministry of Defence, Government of India

13.5 Glimpses of Oil Spills in India

13.5.1 Uran Oil Spill 2013

On 6 October 2013, oil spilled from the Oil and Natural Gas Corporation (ONGC) plant due to a rupture in the Mumbai-Uran Trunk oil pipeline carrying crude oil. Around 5000 litres of crude oil were leaked owing to the rupture, and due to rain, around 1000 to 2000 litres of the leaked oil was mixed with the Arabian Sea for 10 10-kilometer stretch (Petro Online 2013). In pursuance of this, the Ministry of Environment, Forest and Climate Change of India issued a show cause notice to ONGC seeking why action should not be taken against them under the Environment Protection Act, 1986 (Sethi 2013). Additionally, it stated that the pipelines that contributed to the spill should be shut down immediately as per section 5 of the Environment Protection Act, 1986 (Sethi 2013). The seawater was contaminated and due to this, the local communities at large were affected. However, only cleaning up activities were taken up by the corporation, and no further action was taken against ONGC in this regard. A similar incident also happened in the year 2011. A decade later also, a similar incident was reported on September 2023, stating that there was an oil spill owing to the heavy rain, and the authorities controlled the spill and started the cleanup activities. The 2023 oil spill happened in the same pipelines as that of the 2013 oil spills. This depicts that the efforts taken by the corporation, as well as the authorities, are not effective and are based on a case-by-case basis.

13.5.2 M.V. Rak Carrier Oil Spill 2011

In *Samir Mehta vs Union of India*, a ship named M.V. Rak Carrier was crossing Mumbai on 12 August 2011, and it sank due to a technical glitch and water ingress around 20 nautical miles away from the coast (Samir Mehta vs Union of India 2016). M.V. Rak carried around 290 tonnes of fuel oil along with 50 tonnes of diesel that was to be shipped to the Adani Enterprises Limited, dispatched by Delta Group International. Neither the Adani Enterprises Limited nor the Delta Group International took any measures to clean up the spill or control the pollution. Efforts were initiated from the state's side to eradicate the menace, and it sustained huge financial expenditures in doing so. In pursuance of this, a petition was filed by the petitioner Samir Mehta, an environmentalist, before the National Green Tribunal questioning the action of the parties under sections 14 and 15 of the National Green Tribunal Act, 2010, which gives the tribunal the power to settle disputes, and enables the tribunal to order for compensation respectively. The issues raised before the tribunal include questions relating to jurisdiction, power to grant compensation, and liability fixation. Concerning jurisdiction, the tribunal affirmed the existence of jurisdiction by stating that India's sovereignty encompasses resources in contiguous zones and exclusive economic zones under the Maritime Zones Act, 1976, and the tribunal has the power to handle issues involving maritime pollution. The tribunal levied compensation of Rs. 100 crores and Rs. 50 crores are to be submitted to the Ministry of Shipping by Delta Group International and Adani Enterprises Limited, respectively. It is notable that the tribunal has emphasised the principle of no-fault liability and polluters' pay principle while deciding upon the issue.

13.5.3 Ennore Oil Spill

There are several instances of oil spills in Ennore but the most notable one is the collision between two oil tankers in the year 2017, and the most recent one caused by the Michaung cyclone in the year 2023. On 28 January 2017, two tankers, namely BW Maple and Dawn Kanchipuram, collided with each other approximately 2 nautical miles away from the Kamarajar Port. On 4 December 2023, owing to the Michaung cyclone, oil spilled in Ennore Creek and Buckingham Canal. The spill was traced back to Chennai Petroleum Corporation Limited (hereinafter referred to as CPCL). When the matter reached the NGT, it stated that it was not satisfied with the existing cleanup activities, such as deploying fishermen for the cleanup. It also stated that the cleanup activities must be completed by CPCL within 3 days. The Tamil Nadu Pollution Control Board (hereinafter referred to as TNPCB) and CPCL have submitted that 7260 liters of oil have been removed from the region, but the tribunal commented that the authorities have missed the golden hour owing to their delay in action (Venkadesan 2023). As per the report of the TNPCB, 20 tonnes of sand was impacted by the spill and a special task force has been arranged to understand the impact on marine life.

13.6 Economics of Fixation of Liability

The above incidents indicate that the amount of financial loss caused by oil spills is exemplary. The questions that arise in matters of fixation of liability include who will pay the compensation? What will be the amount of compensation? What kind of liability will be fixed? Whether the court will take up the approach of strict liability or absolute liability? As seen earlier in the case of *Samir Mehta v. Union of India*, the Indian courts have time and again emphasised fixing the liability on the polluter by using the polluter's pay principle enshrined in several domestic and international legal agreements. In this case, though the receiver of the goods i.e., Adani Enterprises Limited, did not have any part in the unseaworthiness of the ship or the spill per se, the court imposed no-fault liability on the entity because it was the beneficiary of the transaction under Section 71 of the Merchant Shipping Act, 1958 (Samir Mehta vs Union of India 2016). With respect to the amount of compensation, the courts have differed from case to case basis. The quantum of compensation depends upon the amount of actual loss suffered if the court equates the spill as a tort. The courts can also impose exemplary damages in some cases. The court can also impose a punishment for rash navigation of vessels under the Indian Penal Code, 1860, with simple or rigorous imprisonment for a term which may extend to 6 months, or with a fine which may extend to one thousand rupees, or with both (The Indian Penal Code 1860, § 280). Most of the time the liability that is fixed by the court is civil in nature, but at times they can also impose criminal liability. For example, in the case of *Esso Petroleum Co. Ltd. v. Southport Corporation*, the court rejected the defense of necessity pleaded by the parties and stated that an act of discharging 400 tonnes of oil in the sea to save the crew amounted to a criminal act (Esso Petroleum Co. Ltd. v. Southport Corporation 1956). The courts have also, at times, considered oil spills under tort. "In the case of *Pakkle v. Aiyasami Ganapathi*, the court held that altering the natural quality of water and rendering it less fit for any purpose for which it is capable of being used gives a cause of action under nuisance (Pakkle v. Aiyasami Ganapathi 1956)."

13.7 Conclusion and Suggestions

The oil spill is not only a major environmental concern but also an economic concern, wherein the lives and livelihoods of many who are dependent on the marine ecosystem are affected, as seen in the recent Ennore oil spills in Chennai. Despite having numerous Conventions that have upped their standards to meet the growing concern of oil spills with every incident, the major hurdle that seems to inhibit the effectiveness is the implementation of the procedures involved in the cleanup process and fixing the appropriate compensation proportional to the impact suffered by the victims. Moreover, the cleanup process is mostly done by the residents of the affected area or the fishermen folk, which endangers their lives further owing to the lack of proper equipment and safety gear. Incidents like the Ennore spill, which is not an isolated incident, have highlighted the importance of such implementation, as the event saw fishermen using buckets and rubber gloves to clean up the oil without any systematic process by experienced personnel. This may only clean the surface oil and not the oil that has penetrated the marine life. Further, recognising only the spill from oil tankers and a few kinds of persistent oil limits the scope of such liability.

When fixing the quantum of compensation, certain aspects are to be considered, as the loss of lives, property, and livelihood, with their long-term effects, has to be taken into account. The loss of livelihood doesn't occur only when an oil spill exists. Still, it is also attributed to the loss of employed days owing to the cleanup of their residences and affected properties, loss of revenue due to the refusal to buy fish that might stink of oil,and even the depletion of the marine organisms as an after-effect of the spill. The physical and mental health effects on the victims and their families through generations due to inhalation of the toxins in the oil and related occupational health hazards also need to be included in such computation. While monetary compensation is calculated as per the provisions of law, actual relief to the victims of oil spills is never really provided. Moreover, the laws relating to oil spills do not entirely consider the environmental costs, and in the era of sustainable development, it becomes imperative to do so.

References

Augusyniak CM (1993) Economic valuation of services provided by natural resources: putting a price on the 'priceless'. Baylor Law Rev 45

Coast Guard Act, 1978, § 14

De la Rue C (1997) TOVALOP and CRISTAL – A purpose fulfilled. Int J Shipping Law

DGS (n.d.) Chapter 16: prevention of pollution from ships. Directorate General of Shipping, Government of India. Retrieved from https://www.dgshipping.gov.in/Content/PageUrl.aspx?page_name=ShipManualChap16

Esso Petroleum Co. Ltd. v. Southport Corporation (1956) A.C. 218

Frynas JG (2012) Corporate social responsibility or government regulation? Evidence on oil spill prevention. Ecol Soc 17(4) http://www.jstor.org/stable/26269201

International Maritime Organization (n.d.) Convention on Limitation of Liability for Maritime Claims (LLMC). Retrieved from https://www.imo.org/en/About/Conventions/Pages/Convention-on-Limitation-of-Liability-for-Maritime-Claims-(LLMC).aspx#:~:text=The%20limits%20under%20the%201976,on%20ships%20exceeding%20500%20tons

Jernelöv A (2010) The threats from oil spills: now, then, and in the future. Ambio 39(5/6):353–366. http://www.jstor.org/stable/40801529

Kuokkanen T (2002) International law and the environment. Kluwer Law International

M.V. Elisabeth v. Harwan Investment and Trading (1993) AIR, 1993, 1014

Merchant Shipping Act, 1958, § 10. https://www.dgshipping.gov.in/WriteReadData/userfiles/file/MS%20Act,%201958%20-%20With%20Hyperlinks.pdf

Merchant Shipping Act, 1958, § 365K. https://www.dgshipping.gov.in/WriteReadData/userfiles/file/MS%20Act,%201958%20-%20With%20Hyperlinks.pdf

Mittal R (2023) Wrecks in Indian waters – a Lexi-cultural overview. Blue Waters, Newsletter on Marine Environment Protection 23(1). Retrieved from https://indiancoastguard.gov.in/WriteReadData/NewsLetter/202306140539446277823bluewatersfinal_14Jun23.pdf

NOSDCP, Ministry of Defence, Government of India (2015) National Oil Spill Disaster Contingency Plan. https://indiancoastguard.gov.in/WriteReadData/bookpdf/201512281221565793127NOSDCPCGBR771.pdf

Olawuyi DS (2013) Legal and sustainable development impacts of major oil spills. Consilience 9:1–15. http://www.jstor.org/stable/26476122

Pakkle v. Aiyasami Ganapathi (1956) A.C. 218.

Pearce D, Turner KR (1990) Economics of natural resources and the environment. The Johns Hopkins University Press, Baltimore, MD

Petro Online (2013, October 9) Uran oil leak leaves fishermen seeking compensation. Retrieved from https://www.petro-online.com/news/safety/15/breaking-news/uran-oil-leak-leaves-fishermen-seeking-compensation/27169

Press Information Bureau (2022) Defence Secretary inaugurates two-day National Level Pollution Response Exercise of Indian Coast Guard off Mormugao harbour, Goa. Retrieved from https://pib.gov.in/PressReleasePage.aspx?PRID=1818023

Rathsharma P (2021) Oil spill liability and responses under Indian law: time for an integrated regulatory framework? Indian J Projects, Infrastructure and Energy Law Retrieved from https://ijpiel.com/index.php/2021/09/02/oil-spill-liability-responses-under-indian-law-time-for-an-integrated-regulatory-framework/

Ryland v. Fletcher (1868) L.R. 3 H.L. 330

Samir Mehta v. Union of India (2016) (MANU/GT/0104/2016)

Sangal PS (1995) Indian legal environmental strategy for sustainable development. India Q 51(2/3):33–64. http://www.jstor.org/stable/45072661

Sethi N (2013, October 10) Environment ministry notice to ONGC over Mumbai oil spill. Live mint. Retrieved from https://www.livemint.com/Politics/5JddCGpvGnqigexalPxbJI/Environment-ministry-notice-to-ONGC-over-Mumbai-oil-spill.html

Suleman S (2011) Oil spills: law on liability with special reference to the Indian regime. BLJ 4(1):48. https://ssrn.com/abstract=2044827

The Indian Penal Code, 1860, § 280

Venkadesan (2023, December 15) Chennai: NGT gives 3-day ultimatum to remove oil spill. Times of India Retrieved from https://timesofindia.indiatimes.com/city/chennai/chennai-ngt-gives-3-day-ultimatum-to-remove-oil-spill/articleshow/106006076.cms

Bibliography

Statutes and Conventions

Convention on Limitation of Liability for Maritime Claims (LLMC), 1976

The Coast Guard Act, 1978, No. 30, Acts of Parliament, 1978 (India)

The Indian Penal Code, 1860, No. 45, Acts of Parliament, 1860

The Merchant Shipping Act, 1958, No. 44, Acts of Parliament, 1958 (India)

The Merchant Shipping Amendment Act, No. 12, Acts of Parliament, 1983 (India)

Governmental Reports and Manuals

NOSDCP, Ministry of Defence, Government of India. https://indiancoastguard.gov.in/

Pollution of the Sea by Oil, Chapter 16, Ship Manual. https://www.dgshipping.gov.in/

Mittal R (2023) Wrecks in Indian waters – a Lexi-cultural overview. Blue waters, Newsletter on Marine Environment Protection XXIII(1). https://indiancoastguard.gov.in/

Journal Articles

Augusyniak CM (1993) Economic valuation of services provided by natural resources: putting a price on the 'Priceless'. 45 Baylor L. Rev

De la Rue C (1997) TOVALOP and CRISTAL – a purpose fulfilled. Int J Ship Law

Frynas JG (2012) Corporate social responsibility or government regulation? Evidence on oil spill prevention. Ecol Soc 17(4)

Kuokkanen T (2002) International law and the environment. Kluwer Law International

Olawuyi DS (2013) Legal and sustainable development impacts of major oil spills. Consilience 9:1–15

Pearce D, Turner KR (1990) Economics of natural resources and the environment. The John Hopkins University Press, Baltimore

Rathsharma P (2021) Oil spill liability and responses under Indian law: time for an integrated regulatory framework? Indian J Projects Infrastructure Energy Law, Sep. 02, 2021

Sangal PS (1995) Indian legal environmental strategy for sustainable development. India Q 51(2/3):33–64

E-Journal Articles/Web Sources

Defence Secretary inaugurates two-day National Level Pollution Response Exercise of Indian Coast Guard off Mormugao harbour, Goa, PIB PRESS RELEASE (Apr. 19, 2022, 3:46PM), https://pib.gov.in/PressReleasePage.aspx?PRID=1818023

Jernelöv A (2010) The threats from oil spills: now, then, and in the future. Ambio 39(5/6):353–366. http://www.jstor.org/stable/40801529

Neha Sethi and PTI. Environment ministry notice to ONGC over Mumbai oil spill. MINT (Oct. 10, 2013, 11:47PM), https://www.livemint.com/Politics/5JddCGpvGnqigexalPxbJI/Environment-ministry-notice-to-ONGC-over-Mumbai-oil-spill.html

Suleman S. Oil Spills: Law on Liability with Special Reference to the Indian Regime. 4BLJ 2011(1) at 48. https://ssrn.com/abstract=2044827

Uran oil leak leaves fishermen seeking compensation, PIN online (Oct. 09, 2013). https://www.petro-online.com/news/safety/15/breaking-news/uran-oil-leak-leaves-fishermen-seeking-compensation/27169

Venkadesan S. Chennai: NGT gives 3-day ultimatum to remove oil spill. Times Of India (Dec. 15, 2023, 09:43AM), https://timesofindia.indiatimes.com/city/chennai/chennai-ngt-gives-3-day-ultimatum-to-remove-oil-spill/articleshow/106006076.cms

Case Laws

Esso Petroleum Co. Ltd. v. Southport Corporation (1956) A.C. 218 (India)

M.V. Elisabeth vs Harwan Investment And Trading 1993 AIR 1014 (India)

Pakkle v. Aiyasami Ganapathi, (1956) A.C. 218 (India)

Ryland v. Fletcher (1868), L.R. 3 H.L. 330 (India)

Samir Mehta vs Union of India, MANU/GT/0104/2016 (India)

The author, Ms. R. Vijaya Bharathi, is presently working as an Assistant Professor at the VIT School of Law, Vellore Institute of Technology, Chennai, India. She has been involved in academia for over six years. She obtained her bachelor's degree from School of Excellence in Law, Chennai, and pursued her Master's in Intellectual Property Law from Tamil Nadu National Law University. She is currently pursuing her Ph.D in Fashion Design Piracy and IP as the core area of her doctoral research. She has also qualified for the National Eligibility Test (NET). She has authored book chapters for books published at the national and international levels on themes like outer space laws and tax laws. She has presented papers at various National and International Conferences and Seminars on themes like gender justice, transformative constitutionalism, Artificial Intelligence, and IP. She secured the Best Research Paper Award at a National Conference on Constitutional Law. She co-edited the book "Law and Agriculture: National and International Perspectives" (2022). She has organised several workshops, conferences, and moot court competitions. She has also participated in various Faculty Development Programs. Her areas of interest include IPR, gender studies, and outer space laws.

The author, Ms. S. R. Ramya is an ardent learner of life. She is presently working as an Assistant Professor in the Department of Taxation Law at the Tamil Nadu Dr. Ambedkar Law University (TNDALU), Chennai, India. She is a Doctoral Research Scholar at the West Bengal National University of Juridical Sciences (NUJS), India. She is an engineer-turned-lawyer who has been involved in academia for almost four years. Having done her Master's in Taxation Law, she is currently pursuing her Ph.D with Transfer Pricing as the core area of her doctoral research. She qualified for the UGC- NET and JRF in her first attempt in 2019. She has authored around 15 research articles published in esteemed international, refereed journals, as well as book chapters, and presented papers at National and International Conferences and Seminars, including one at HPNLU and at VSL and the University of Birmingham, on themes such as gender neutrality, sexual healthcare, and comparative tax laws. She has also participated in various Faculty Development Programs and expert conclaves. She has successfully organized National-level seminars/ conferences, outreach programmes including visits to Parliament and State budget sessions, and adjudicated as the chair in various National and International Moot court competitions. Her other academic and research interests include sports law and gender studies.

Chapter 14
Industrial Jurisprudence and Labour Legislation: A Journey of Efforts to Conquer Income Disparity

Nisha Jindal and Renu Sharma

Abstract Scarcity is always the result of high **demand** against a limited supply. Similarly, inequality in the industrialized world is the result of rising demand for skilled workers bordering on a surfeit of unskilled workers. This imbalance creates a policy dilemma for nations striving to reduce inequality without stifling economic growth. The paper traces the evolution of labour legislation and examines how this jurisprudence contributes to income disparity. It also examines the development of Industrial policies from early independence to the globalization era of 1991. The primary focus is on understanding how income disparities affect different social groups, particularly workers, and on examining the legal frameworks and mechanisms necessary to address these inequalities.

Keywords Labour laws · Disparities · Inequality of incomes · Inequality of opportunities · Consumption · Wealth · Labour market · Wage inequality

14.1 Introduction

Social inequalities are evident in every country. Differences in wealth, material possessions, power, prestige, job accessibility, financial resources, social services, and basic necessities like food, shelter, and drinking water significantly influence the social dynamics of contemporary societies. While inequalities take many forms,

---

The chapters in this book are presented in the "International Conference on Legal and Economic Synergy for an Equitable Society, organised by the University Institute of Laws at Panjab University Regional Centre, Ludhiana, India on February 1st, 2024".

N. Jindal (✉) · R. Sharma
University Institute of Laws, Panjab University Regional Centre, Ludhiana, India
e-mail: nishajindal4@gmail.com; renusharma804@gmail.com

A. Virk et al. (eds.), *Law, Economy, and Society: Insights into Legal and Economic Issues in India*, Accounting, Finance, Sustainability, Governance & Fraud: Theory and Application, https://doi.org/10.1007/978-981-95-3361-9_14

economic disparity, social status, and power are the most impactful on access to income opportunities, their distribution, and consumption patterns. Throughout the twentieth century, especially after World War II, many societies have endeavored to reduce social inequalities stemming from disparities in wealth and social status and to address the persistent effects of discrimination on certain social groups (UN 2001).

Contributions of Labour are evident in all societal sectors, leading to the country's development. However, wage and earnings inequality persist not only in the unorganized sector but also in the organized sector. To create a more egalitarian society, it is necessary to invest in social infrastructure, healthcare, education, affirmative action, and public services.

Development cannot be discussed without addressing income inequality, and it is a recent topic that caught the attention of the IMF, World Bank, OECD, and Davos meetings (Dev 2018). For the first time globally, the reduction of inequality is included as a Sustainable Development Goal (SDG) under Goal 10. By 2030, Target 1 aims to achieve an income growth rate for the bottom 40% that is higher than the national average, and Target 2 seeks to promote the socio-economic and political inclusion of all.

14.2 Historical Overview

Labour Laws were not enacted in one fell swoop but evolved with the needs of labourers. Steadily growing from the industrial revolution of Great Britain in the eighteenth century, when the factory system upgraded the economy and workers migrated to industrial towns, goods were produced at a mass scale and at cheaper rates, leading to a shift from handicraft to machine production (Misra 2008).

A similar wave was seen in British India, where the motto was to earn profits at the cost of forced labor, in which the latter had to work like machines for long hours without any rest. As a result of labor strikes, the British government passed several laws, namely the Indian Slavery Act 1843; the Workmen's Compensation Act 1923, the Indian Trade Unions Act 1923, the Payment of Wages Act 1936, and finally the Industrial Employment Standing Orders Act 1946 (Bose and Bose 2022).

Making a difference, the Constitution of Independent India was inspired by the vision of nationalist leaders, the debates of the Constituent Assembly, and the provisions of International conventions. Its Fundamental Rights and DPSP highlighted the need to safeguard workers' interests and emphasised labour's dignity.[1] Additionally, numerous laws were passed, including the Industrial Dispute Act 1947, the Minimum Wages Act and the Factories Act of 1948, and the Equal Remuneration Act 1976, etc.[2] However, India has also formulated the policies for Industry which are as under:-

[1] https://lawbhoomi.com/history-of-labour-law-in-india (accessed on 21.01.2024).

[2] Supra note 7.

14.2.1 The Classical Framework: Nehru Mahalanobis Model

In line with the vision of the top leadership, India's Five-Year Plans were crafted to foster economic and social development within a 'socialist' framework. These plans targeted multiple goals, namely industrialization, an increase in per capita income, and ensuring equitable distribution of the benefits from economic growth. Additionally, they aimed to reduce the concentration of economic power and promote more balanced regional industrial development. The key elements of the economic strategy during the 1950s, 1960s, and much of the 1970s included (Singh 2008):

- Emphasizing the role of heavy industry in economic development and rapidly building the capital goods sector.
- Assigning a leading role to the public sector in the economy's structural transformation.
- Guiding major private sector investments based on the requirements of the national plan, rather than private profitability.
- Prioritizing technological self-reliance, with a strong inward orientation, meaning that domestically producible goods should not be imported, regardless of the cost.

P.C. Mahalanobis advocated for a capital goods-focused industrial strategy in India, viewing the development of this sector as vital for economic growth. While suitable for a closed economy, this approach could also be justified for India, considering factors like slow global growth, falling commodity prices, and advantageous industries such as steel with nearby raw materials. However, this strategy clashed with the 5-year plans' employment objectives. To reconcile, plans shielded small-scale, cottage industries from competition, limiting the expansion of modern factories to prevent clashes with costly industry goods.

The Indian government employed various measures to execute its industrial strategy and synchronize the private sector with national goals. These measures encompassed industrial licensing for new products and capacity expansion, rigorous import controls, export subsidies, price regulation and stringent oversight of multinational investments. Jawaharlal Nehru played a key role in founding institutions focused on scientific and technical infrastructure, paving the way for the subsequent development of the information and communications technology sector.

14.2.2 Economic Reforms of 1991: LPG Model

Soon after the disintegration of the USSR, the Government of India too announced the long-awaited liberalised industrial policy in 1991, amidst severe economic instability. The objective was to raise efficiency and accelerate economic growth. LPG Model (Liberalisation, Globalisation and Privatisation) brought significant changes, including reducing public sector reservation to only two sectors (atomic energy and railways), abolishing licensing except for specific industries related to security concerns, and disinvesting government stakes in Public Sector Enterprises to improve efficiency. It also liberalized foreign investment, allowing majority stakes in many industries and up to 100% FDI in various sectors.

The manufacturing sector in India has shown stagnation, with its contribution to GDP remaining at about 16% since 1991 despite industrial policies. The selective inflow of investments has led to distortions in the industrial pattern, with significant investments in some industries but slow progress in essential sectors like engineering, power, and machine tools. Additionally, there is a lack of a well-defined industrial location policy in the New Industrial Policy, which could ensure environmentally sustainable industrial development.

Major impact was on the working class of India. Laborers were displaced due to the restructuring and modernization of industries following the new industrial policy. The lack of incentives to improve efficiency resulted from a focus on internal liberalization without sufficient emphasis on trade policy reforms, leading to a growth model driven more by consumption than by investment or exports.

14.2.3 National Manufacturing Policy, 2011

After the global financial crisis of 2008, there was wide consensus among policymakers on the need for an enhanced role of the government in facilitating sustainable industrial growth and employment creation.

India's economic success had largely been driven by growth in the services sector, while the industrial sector, particularly manufacturing, stagnated despite strong policy measures. The manufacturing sector's contribution to GDP was around 15%, significantly lower than other emerging economies like China, where it is close to 45%. This stagnation prevented India from providing sufficient employment for its large labor force. The lack of manufacturing jobs placed undue pressure on the agriculture sector to absorb the workforce, which is not feasible under any economic model. This led to the phenomenon of "Jobless Growth" unique to India. Recognizing this issue, the government launched the National Manufacturing Policy in November 2011 to promote the manufacturing sector.

14.2.4 Make in India, 2015

The Make in India initiative, launched in 2014, marks a shift from India's earlier policy of self-sufficiency from the 1970s. Unlike the previous approach, it moves away from the licence raj, self-sufficiency, and import-substituting industrialization. The initiative aims to boost domestic manufacturing and transform India into a global manufacturing hub by facilitating investment, fostering innovation, enhancing skill development, protecting intellectual property, and building world-class manufacturing infrastructure.

14.2.5 PLI Scheme 2020: Indian Policy Beginning to Mature

Following the launch of the "Make in India" initiative, there was significant excitement about India potentially developing a competitive manufacturing base. However, it took the government 6 years to develop a substantial industrialization strategy. With the launch of Production Linked Incentives (PLI) Scheme[3] in March 2020, India not only took a consequential step towards exports-centric manufacturing but also kick-started its own industrial policy making. Now, over the past year, there are clear signs suggesting that India's industrial policy is not just working, but also beginning to mature.[4]

14.3 Contemporary Labour Legislation

The term 'labour' refers to 'productive work', particularly physical work done for wages. Labour law, commonly called Employment law, is a collection of laws, administrative rulings, and precedents that define legal rights and restrictions of workers and their organizations. Labour Laws comprise two main categories: Collective and Individual. Former deals with the tripartite relationship of employees, employers, and unions. Latter focuses on employees' rights at work and through their employment contracts.[5]

On the basis of specific objectives which it has sought to achieve, the labour legislations can be classified into the following categories:[6]

1. *Regulative Legislation*—Primary goal is to manage the relationships between employees and employers and to establish procedures for resolving industrial disputes. These laws also oversee the interactions between workers and their trade unions, define the rights and responsibilities of employer and worker organizations, and regulate their mutual relationships.
2. *Protective Legislation*—Aimed at protecting labor standards and enhancing working conditions. These laws set minimum standards for areas such as working

---

[3] https://www.investindia.gov.in/production-linked-incentives-schemes-india (accessed on 25.04.2024)

[4] https://www.orfonline.org/research/india-s-industrial-policy-is-maturing (accessed on 25.04.2024)

[5] Labour Laws in India https://ncib.in/pdf/ncib_pdf/Labour%20Act.pdf (accessed on 25.04.2024).

[6] Unit 3 Labour Laws - Concept, Origin, Objectives & Classification (accessed on 25.04.2024).

hours, the employment of children and women, and labor conditions in factories, mines, plantations, transportation, shops, and other establishments.

3. *Wage-Related Legislation*—Lays down the methods and manner of wage payment as well as the minimum wages.
4. *Social Security Legislation*—Intend to provide workmen's social security benefits under certain contingencies of life and work.
5. *Welfare Legislation* (both inside and outside Workplace)—Aim is to enhance the general welfare and living conditions of workers. While all labor laws can be seen as promoting worker welfare and improving living conditions, and many protective labor laws include provisions on labor welfare, the laws in this category are specifically focused on improving the living conditions of workers.

Listed under Concurrent List of Indian Constitution, both Union and State legislatures have framed 45 and 170 statutes respectively, to govern the labour market. These laws have recently been subsumed under four codes:[7]

- ***Code on Wages*** (consolidates Payment of Wages Act 1936, Minimum Wages Act 1948, Payment of Bonus Act 1965, and Equal Remuneration Act 1976)
- ***Code on Industrial Relations*** (deals with strikes, lockouts and retrenchments, consolidates Trade Unions Act 1926, Industrial Employment Act 1946, Industrial Disputes Act 1947)
- ***Code on Health, Safety and Working Conditions*** (replaces 13 previous laws including Factories Act 1948, Mines Act 1952, Contract Labour Act 1970)
- ***Code on Social Security*** (replaces multiple laws, including Compensation Act 1923, Employees State Insurance Act 1948, Employees Provident Fund Act 1952, Maternity Benefit Act 1961, Payment of Gratuity Act 1972, Cine Workers Welfare Fund Act 1981, Unorganised Workers Social Security Act 2008).

14.4 Social: Equity and Security

Every being has a different perspective of Social Justice being guaranteed by the Indian Constitution.[8] In the words of Dr. B.R. Ambedkar, a just society is one in which an ascending sense of reverence and a descending sense of contempt is dissolved into the creation of a compassionate society. Social Justice is lacking when there is a gap between owners of wealth and property and those excluded and deprived. Principle to treat like cases alike and unlike differently is apt; it doesn't signify equality in all means but suggests that inequality beyond imagination is a threat to social justice. NITI Aayog's objective is in consonance with the same, stating that every being must enjoy the fruits of development to achieve social justice.

[7] Ibid.

[8] See Part III and Part IV of Indian Constitution

The concept of social security has become an integral part of social justice. It is based on ideas of human dignity and social justice. The underlying idea behind social security measures is that a citizen who has contributed or is likely to contribute to his country's welfare should be given protection against certain hazards.[9] Its measures are significant from two viewpoints, i.e. they constitute an important step towards the goal of a welfare state; and secondly, they enable workers to become more efficient and thus reduce wastage arising from industrial disputes. Lack of social security impedes production and prevents formation of a stable and efficient labour force. Therefore, social security measures are not a burden but a wise investment which yields good dividends.[10]

In 1952, the International Labour Organisation adopted the Social Security (Minimum Standards) Convention, which embodies universally accepted basic principles and common standards of social security. The application of these principles has guarded developments of this field throughout the world.[11] Accordingly, in India, a number of social security legislations have been enacted from time to time to promote the condition of the labour, keeping in view the development of industry and national economy. Some of these enactments are:

- The Workmen's Compensation Act, 1923
- The Trade Union Act, 1926
- The Industrial Employment (Standing Orders) Act, 1946
- The Industrial Disputes Act, 1947
- The Minimum Wages Act, 1948
- The Employee's State Insurance Act, 1948
- The Factories Act, 1948
- Maternity Benefits Act, 1961

Moreover, it has been observed by the Hon'ble Supreme Court of India that Social Security for the more fragile sections is of utmost significance. Be that as it may, at that point we can't overlook the limitations under which we are living. While we should not overlook our social goals, our purpose might be vanquished on the off chance that we don't move toward our problems in a common-sense manner. Thus, social justice requires the Industrial Courts to strike a harmony between the clashing claims of employer and employee. While the employer has a key option to maintain his industry, his privilege has to be adjusted with the worker's entitlement to social justice.[12]

---

[9] I.L.O. Approaches to Social Security 1942, p. 80, as quoted in Report of the National Commission on Labour, 1969, p. 162.

[10] V.V.Giri, Labour Problems in Indian Industry, p. 248.

[11] Report of the National Commission of Labour 1969, p. 162.

[12] Patna Electric Supply v. Workers Union, 1959 AIR 1035.

14.5 Labor Institutions and Inclusive Development

Although belief of 'Development' differs for individuals as being positive for one might be destructive for another, yet a crucial element to measure economic development is industrial productivity which in turn needs labour productivity. Labour as an activity involves both the ones who provide capital and also those putting physical efforts in the production process.

Connecting Ishikawa Model with recent models for measuring Quality of Working Life, we found a set of elements to be managed by the production system to have the best of workers. Ishikawa relates 'organisational performance' to working life quality and the model depicted below applies to each organisation, including industry.[13]

Knowledge × Skill = Ability
Attitude × Environment = Motivation
Ability × Motivation = *Individual performance*
Individual Performance × Organisational resources = *Organisational performance*
Organisational Performance × Society = Quality of Working Life

It was argued in the V.V. Giri Memorial Lecture delivered at Indian Society of Labour Economics, JNU Delhi, that labour law should be seen as a development institution, capable of promoting both equality and efficiency, leading to inclusive development.[14] The link between labour institutions and economic development is examined at two levels. At the micro level, it examines socio-economic forces resulting in individuals' productive work and their working conditions. At the macro level, the objective is to comprehend implications for economic growth (efficiency) and distribution (equity).[15] The primary focus of the contemporary world is that Economic development and social progress should grow side by side.

14.6 Income Disparities

Addressing economic disparities is the biggest challenge in today's world. Irrespective of ideology, culture, and religion, people care about inequality as it is a reflection of persistent disadvantage to particular segments of society (Era Dabla-Norris 2015).

---

[13] S.P. Mukherjee, Labour: The Lever of Development, (V.V. Giri National Labour Institute, U.P. 2019).

[14] Ibid.

[15] K P Khanna, Labour Institutions and Economic Development in India. https://www.indianlabourarchives.org/bitstream/20.500.14121/1815/1/Labour%20Institutions%20and%20Economic%20Development%20in%20India%20Some%20Exploratory%20Hypotheses.pdf (accessed on 18.2.24).

The GINI Coefficient published by the World Bank is an accepted measure of income inequality. It measures the dispersion of income across the frequency distribution of income. The higher the number, the larger the inequality (Gosh 2022). Highly unequal societies are slower in economic growth and less efficient at poverty reduction, as compared with those of lower inequality levels. Disparities in health and education forces them into a vicious circle, transmitting disadvantage from one generation to next.[16]

Multiple theorists explain the concept differently. According to the World Development Report of 2006, even basic opportunities for life are inequitably distributed. These disparities raise political tension, which further creates an inefficient and damaged social system, favouring those with more influence and affluence. As a result, the country misses the opportunities for innovation and invention.[17]

Recently, India too has joined the club of the most unequal countries. As per the income equality data of India Human Development Survey, India scores lower than Russia, US, China and Brazil, and is more egalitarian than South Africa only. Moreover, according to Credit Suisse, the richest 1% Indians own 53% of the country's wealth, and the poorer half hold a mere 4.1% of national wealth. The increase in income inequality coincides with the sharp rise in Indian economic growth after 1980.[18]

14.7 Causes of Income Disparity in India

A new study[19] from the World Inequality Lab finds that the present-day golden era of Indian billionaires has produced soaring income inequality in India—now among the highest in the world and starker than in the U.S., Brazil, and South Africa. The gap between India's rich and poor is now so wide that by some measures, the distribution of income in India was more equitable under British colonial rule than it is now, according to the group of economists who co-authored the study, including the renowned French economist Thomas Piketty.

The current total number of billionaires in India is peaking at 271, with 94 new billionaires added in 2023 alone, according to Hurun Research Institute's 2024 global rich list[20] published Tuesday. That's more new billionaires than in any country

[16] World social report 2020: Inequality in a rapidly Changing World, United Nations: Department of Economic and Social Affairs.

[17] A.Banerjee, P.K. Kuri, Development Disparities in India, India Studies in Business and Economics (Springer India 2015).

[18] https://www.civilsdaily.com/income-inequality-in-india-causes-remedies-and-consequences/

[19] INCOME AND WEALTH INEQUALITY IN INDIA, 1922–2023 https://wid.world/wp-content/uploads/2024/03/WorldInequalityLab_WP2024_09_Income-and-Wealth-Inequality-in-India-1922-2023_Final.pdf (accessed on 27.04.2024).

[20] Info - Hurun Global Rich List 2024 https://www.hurun.net/en-US/Info/Detail?num=K851WM942LBU (accessed on 27.04.2024).

other than the U.S., with a collective wealth that amounts to nearly $1 trillion—or 7% of the world's total wealth. Authors write, *"The Billionaire Raj headed by India's modern bourgeoisie is now more unequal than the British Raj headed by the colonialist forces"*.[21]

Main causes for these Income disparities are as follows:

14.7.1 Globalization

Globalization played a major role in creating unemployment by shifting focus from agriculture to the manufacturing sector, which requires skilled labour. People are losing their jobs to foreigners who are willing to provide services at lower wages. For instance, in the USA, wages have decreased significantly due to immigration.[22]

14.7.2 Regional Inter-State Inequalities

Regional differences in living standards have implications for economic and political functioning of national economies. Such disparities lead to undesirable exchange of labour-capital and inflammation between depressed and prosperous regions. It worsens the aggregate trade-off between inflation and unemployment.[23] For instance, Mumbai, where 233 billionaires live, is home to millions of slums. Another example of Bihar, having lowest rung in India's per capita income, had the richest candidate for 2019 Lok Sabha elections having assets worth Rs. 1107 Crore. These shows emerging dimensions of inequalities within states of India.[24]

---

[21] India's Income Inequality Worse Than Under British Rule: Report I TIME https://time.com/6961171/india-british-rule-income-inequality/ (accessed on 27.04.2024).

[22] The Effects of Globalization to Employment and International Trade I Free Essay Example (studycorgi.com) StudyCorgi. (2022, April 15). *The Effects of Globalization to Employment and International Trade.* Retrieved from https://studycorgi.com/the-effects-of-globalization-to-employment-and-international-trade/

[23] A.P.thirlwall, Regional Disparities in Per capita Income in India: Convergence or Divergence? (School of Economics Discussion Papers, No. 1313, University of Kent, School of Economics, Canterbury) August 2013. https://www.econstor.eu/bitstream/10419/105585/1/1313.pdf

[24] Regional Inequality in India: A State Level Analysis Pandey, Aviral and Gautam, Richa A N Sinha Institute of Social Studies, Patna, M.R.M.College, Lalit Narayan Mithila University, Darbhanga 2020. https://mpra.ub.unimuenchen.de/101980/1/MPRA_paper_101980.pdf. (accessed on 15.02.2024).

14.7.3 Lack of Productive Jobs for the Poor

For Growth, investment is a requisite, and its dearth limits the productive capacity of different sectors, leading to inadequate generation of employment opportunities in the economy.[25] Indian economy is primarily based on the slowest growing agrarian sector that stands in need of investment, whereas industrial investments can create productive jobs for the poor. To state an example of Bihar, which has achieved higher growth in overall state's income, a large chunk of its population is not getting benefited as they are dependent on agriculture. Thus, it can be said that economic growth in Bihar has been less inclusive than in India as a whole.[26]

14.7.4 Gender Gaps

India resides 121 crore people, among whom around 48% are females, but continue to be underrepresented at the level of politics and governance. One can easily find inconsistency in opportunities of female and male population in India. As per International Labour Organization Report of 2019, women are paid 34% less than men, this gender wage gap in India is highest among all 73 countries studied.

That is why the concept of women empowerment evolved and policies were designed such that they can get equal opportunities to excel their life.[27] Even after five decades of planned economic development, we are unable to provide economic independence by providing enough gainful employment to the female workforce.[28]

14.7.5 Geographical Dimensions

Income inequality also depends upon the geographic structure of the state. The difficult terrain surrounded by hills, rivers, and dense forest leads to increased cost of administration and development projects. Also adverse climate and floods are responsible factors for the lower rate of economic development of certain regions as reflected by low agricultural productivity and lack of industrialization.[29]

---

[25] THE PROBLEM OF UNEMPLOYMENT, POVERTY AND INEQUALITY https://nios.ac.in/media/documents/SrSec318NEW/318_Economics_Eng/318_Economics_Eng_Lesson4.pdf

[26] /https://mpra.ub.unimuenchen.de/101980/1/MPRA_paper_101980.pdf

[27] Supra note 23.

[28] M.Koteswara Rao, Empowerment of Women in India 131 (Discovery Publishing House, Delhi, 2005).

[29] Supra note 24.

14.7.6 *Migration of Labourers*

Migration is a shift from residence, either fora specified duration or permanently.[30] Recent migration data released by Census shows the emerging situation of livelihood crisis faced by households in some states. According to the 2011 Census, 'Hindi Belt' comprising Uttar Pradesh, Bihar, Rajasthan, and Madhya Pradesh accounts for 50% of India's total inter-state migrants. Major destinations of these states are Delhi, Maharashtra, Tamil Nadu, Gujarat, Andhra Pradesh, and Kerala.[31] Major reasons for such migration is job search and non-availability of economic opportunities in their place.

Economic diversification and increased land productivity, rapid improvement in transport and communication means, improvement in education and zeal for improving living standards, have added momentum to the mobility of population in India (Roy 2011).[32]

14.7.7 *Lack of Skill Development Programs*

Skills Gap in India is significant as the Indian Education System provides zero professional education, and students after passing high school have to get certain vocational training to get a job. As per the National Council for Skill Development and Entrepreneurship (NCSDE) Report, 53% of Indian businesses were unable to recruit in 2009 because of skills shortage. In spite of India having a younger population than most countries, its schools fail to provide a useful curriculum for the modern job market (Chanda and Bandyopadhyay 2023).

14.8 Conclusion

It is clear that laborers are essential to development; yet, income and employment disparities persist due to various factors. Industrialization, necessary for job creation and societal advancement, typically results in industries being established in locations where raw materials are affordable and accessible. This leads to increased labor migration and economic inequality between regions.

Effective labor legislation, adapted to the modern world's economic and social challenges, plays three critical roles. Firstly, it establishes a legal framework that

[30] Causes and Impact of Labour Migration: A Case Study of Punjab Agriculture Baljinder Kaur*, J.M. Singh, B.R. Garg, Jasdev Singh and Satwinder Singh, Agricultural Economics Research Review Vol. 24 (Conference Number) 2011 pp. 459.

[31] Supra note 23.

[32] Supra note 29.

fosters productive individual and collective employment relationships. Secondly, it provides a structured environment for employers, workers, and their representatives to address work-related issues. Thirdly, it guarantees and consistently upholds fundamental work principles and rights that are widely accepted in society, while also creating mechanisms for their implementation and enforcement.[33]

The paper presents several recommendations to reduce inequality among laborers. It advocates for promoting labor-intensive manufacturing to create more jobs and absorb surplus labor. Introducing new technology and equipment in agriculture is suggested to enhance productivity and prevent the migration of workers to other industries. Implementing skill development programs is also emphasized to provide better job opportunities for workers. The paper calls for progressive taxation on the wealthy and luxury items to ensure a more equitable distribution of wealth. It highlights the importance of government initiatives to provide equal employment opportunities for all and encourages the employment of women across various sectors to reduce income disparities within households.

References

Banerjee A, Kuri PK (2015) Development disparities in India, India studies in business and economics. Springer India

Bose SP, Bose J (2022) Introduction to labour Laws in India. https://www.google.co.in/books/edition/Introduction_to_Labour_Laws_in_India. Accessed 25.1.24

Baljinder K, Singh JM, Garg BR, Singh J, Singh S (2011) Causes and impact of labour migration: a case study of Punjab Agriculture. Agric Econ Res Rev 24 (Conference Number):459

Chanda A, Bandyopadhyay S (2023) A wise book Gurcharan Das Work3.0. Penguin Random House India

Dev SM (2018) Inequality, employment & public policy. Indira Gandhi Institute of Development Research, Mumbai. http://www.igidr.ac.in/pdf/publication/WP-2018-003.pdf. Accessed 19.12.23

Era Dabla-Norris, Kochhar K, Suphaphiphat N, Ricka F, Tsounta E (2015) Causes and consequences of income inequality: a global perspective. Staff Discussion Notes 2015, 013. Accessed 15 Aug 2025

Ghosh S (2022) Indian economy, 151. PHI Learning Private Limited, Delhi

Giri VV. Labour Problems in Indian Industry, p 248

https://www.imf.org/en/Publications/Staff-Discussion-Notes/Issues/2016/12/31/Causes-and-Consequences-of-Income-Inequality-A-Global-Perspective-42986

https://www.investindia.gov.in/production-linked-incentives-schemes-india. Accessed 25.04.2024

https://www.orfonline.org/research/india-s-industrial-policy-is-maturing. Accessed 25.04.2024

I.L.O. Approaches to Social Security (1942), p. 80, as quoted in Report of the National Commission on Labour, 1969, p 162

Income And Wealth Inequality In India 1922–2023. https://wid.world/wpcontent/uploads/2024/03/WorldInequalityLab_WP2024_09_Income-and-Wealth-Inequality-in-India-1922-2023_Final.pdf. Accessed 27.04.2024

India's Income Inequality Worse Than Under British Rule: Report | TIME. https://time.com/6961171/india-british-rule-income-inequality/. Accessed 27.04.2024

[33] Supra note 10.

Inequality, Employment & Public Policy. http://www.igidr.ac.in/pdf/publication/WP-2018-003.pdf. Accessed 19.12.2023

Info-Hurun Global Rich List 2024. https://www.hurun.net/en-US/Info/Detail?num=K851WM942LBU. Accessed 27.04.2024

Khanna KP. Labour Institutions and Economic Development in India. https://www.indianlabourarchives.org/bitstream/20.500.14121/1815/1/Labour%20Institutions%20and%20Economic%20Development%20in%20India%20Some%20Exploratory%20H potheses.pdf. Accessed 18.2.2024

Koteswara Rao M (2005) Empowerment of women in India 131. Discovery Publishing House, Delhi

Labour Laws in India. https://ncib.in/pdf/ncib_pdf/Labour%20Act.pdf. Accessed 25.04.2024

Misra SN (2008) Labour & Industrial Laws, 4. Central Law Publication, New Delhi

Mukherjee SP (2019) Labour: the lever of development. V.V. Giri National Labour Institute, Uttar Pradesh

Part III and Part IV of the Indian Constitution

Patna Electric Supply v. Workers Union, 1959 AIR 1035

Regional Inequality in India (2020) A state level analysis Pandey, Aviral and Gautam, Richa A N Sinha Institute of Social Studies, Patna, M.R.M.College, Lalit Narayan Mithila University, Darbhanga

Report of the National Commission of Labour (1969) p 162

Singh A (2008) The past, present and future of industrial policy in India: Adapting to The Changing Domestic and International Environment. https://www.jbs.cam.ac.uk/wp-content/uploads/2023/05/cbrwp376.pdf. Accessed 15 March 2025

Siva Prasad Bose &Joy Bose, Introduction to Labour Laws in India. https://www.google.co.in/books/edition/Introduction_to_Labour_Laws_in_India. Accessed 25.1.2023

The Effects of Globalization on Employment and International Trade | Free Essay Example (studycorgi.com) StudyCorgi (2022, April 15) The Effects of Globalization on Employment and International Trade. Retrieved from https://studycorgi.com/the-effects-of-globalization-to-employment-and-international-trade/

The Past, Present and Future of Industrial Policy In India: Adapting To The Changing Domestic and International Environment, n.d.. https://www.jbs.cam.ac.uk/wp-content/uploads/2023/05/cbrwp376.pdf

The Problem Of Unemployment, Poverty And Inequality Https://Nios.Ac.İn/Media/Documents/Srsec318new/318_Economics_Eng/318_Economics_Eng_Lesson4.Pdf

Thirlwall AP (2013) Regional Disparities in Per capita Income in India: Convergence or Divergence? (School of Economics Discussion Papers, No. 1313, University of Kent, School of Economics, Canterbury) August 2013. https://www.econstor.eu/bitstream/10419/105585/1/1313.pdf

UN (2001) Disparities in incomes and poverty, Chapter XI. https://www.un.org/esa/socdev/rwss/docs/2001/11%20Disparities%20in%20incomes%20and%20poverty.pdf. Accessed 18.1.24

Unit 3 Labour Laws - Concept, Origin, Objectives & Classification. Accessed 25.04.2024

World social report (2020) Inequality in a rapidly changing world, United Nations: Department of Economic and Social Affairs

Dr. Nisha Jindal (LL.B., LL.M., Ph.D.), is an Assistant Professor at the University Institute of Laws, Panjab University Regional Centre, Ludhiana, with a remarkable academic background and extensive experience in teaching, research, and law practice. A gold medalist from Panjab University, Dr. Jindal holds a Ph.D. in Copyright Laws from Maharaja Agrasen University. Dr. Jindal has also successfully cleared the UGC NET examination. She has actively participated in numerous seminars, conferences, and workshops, showcasing her expertise in the field. Dr. Jindal has been invited as a judge in various competitions and has delivered guest lectures on esteemed platforms. She was awarded the best contributor for the moot court competition by the Association of World Education Events. Prior to her academic career, she practiced as an Advocate at the Hon'ble Punjab & Haryana High Court, Chandigarh, and District Court, Ludhiana, handling critical cases with proficiency.

Dr. Renu Sharma (LL.B., LL.M., Ph.D.) is serving as a Part-time Assistant Professor at the University Institute of Laws, Panjab University Regional Centre, Ludhiana. With a Ph.D. from Patiala, Dr. Sharma has established herself as an expert in her field, actively participating in numerous seminars, conferences, and workshops. Her extensive legal experience includes practicing as an Associate Advocate at the District Courts of Bathinda and Ludhiana, as well as working as a Legal Consultant in Delhi's Tees Hazari Courts, showcasing her expertise and commitment to the legal profession.

Correction to: Surrogate Advertising in India

Aman Amrit Cheema and Ankur Taya

Correction to:
Chapter 5 in: A. Virk et al. (eds.), *Law, Economy, and Society: Insights into Legal and Economic Issues in India*, Accounting, Finance, Sustainability, Governance & Fraud: Theory and Application, https://doi.org/10.1007/978-981-95-3361-9_5

The original version of the chapter was inadvertently published with incorrect affiliation of the author Ankur Taya. The correct affiliation details have been updated in the chapter.

The updated version of this chapter can be found at
https://doi.org/10.1007/978-981-95-3361-9_5

A. Virk et al. (eds.), *Law, Economy, and Society: Insights into Legal and Economic Issues in India*, Accounting, Finance, Sustainability, Governance & Fraud: Theory and Application, https://doi.org/10.1007/978-981-95-3361-9_15

Index

A. Virk et al. (eds.), *Law, Economy, and Society: Insights into Legal and Economic Issues in India*, Accounting, Finance, Sustainability, Governance & Fraud: Theory and Application, https://doi.org/10.1007/978-981-95-3361-9

MIX
Papier aus verantwortungsvollen Quellen
Paper from responsible sources
FSC® C105338

If you have any concerns about our products,
you can contact us on
ProductSafety@springernature.com

In case Publisher is established outside the EU,
the EU authorized representative is:
Springer Nature Customer Service Center GmbH
Europaplatz 3, 69115 Heidelberg, Germany

Printed by Libri Plureos GmbH
in Hamburg, Germany